MEMS and Microfluidic Devices for Analytical Chemistry and Biosensing

MEMS and Microfluidic Devices for Analytical Chemistry and Biosensing

Editor

Stefano Zampolli

MDPI • Basel • Beijing • Wuhan • Barcelona • Belgrade • Manchester • Tokyo • Cluj • Tianjin

Editor
Stefano Zampolli
IMM-CNR Institute for Microelectronics and
Microsystems—Italian National Research Council
Italy

Editorial Office
MDPI
St. Alban-Anlage 66
4052 Basel, Switzerland

This is a reprint of articles from the Special Issue published online in the open access journal *Micromachines* (ISSN 2072-666X) (available at: https://www.mdpi.com/journal/micromachines/special_issues/mems_microfluidic_devices_analytical_chemistry_biosensing).

For citation purposes, cite each article independently as indicated on the article page online and as indicated below:

LastName, A.A.; LastName, B.B.; LastName, C.C. Article Title. *Journal Name* **Year**, *Volume Number*, Page Range.

ISBN 978-3-0365-6201-8 (Hbk)
ISBN 978-3-0365-6202-5 (PDF)

Contents

About the Editor

Stefano Zampolli

Stefano Zampolli (MPhys) graduated in physics at the University of Bologna in 2000, producing a thesis on the application of Fuzzy Logic for the detection of buried landmines. Since April 2000, he has been a grant student at CNR-IMM Bologna, working primarily on gas sensor characterization and data processing within the sensor and microsystem R&D program. Since 2005, he has been a researcher in the sensors and microsystem group at CNR-IMM, designing and developing gas sensors and gas sensing microsystems for environmental monitoring, safety and security, and agrofood applications.

micromachines

MDPI

Editorial

Editorial for the Special Issue on MEMS and Microfluidic Devices for Analytical Chemistry and Biosensing

Stefano Zampolli

CNR-IMM Institute for Microelectronics and Microsystems, Italian National Research Council, Via P. Gobetti 101, 40129 Bologna, Italy; zampolli@bo.imm.cnr.it

The outbreak of the SARS-CoV-2 pandemic has made the general public aware of the breakthrough technologies which were developed in recent years for state-of-the-art biosensing, and terms such as clinical specificity and sensitivity are now widely understood. The need for reliable point-of-care diagnostic systems has never been felt as evidently as during the last few years. While PCR and other molecular diagnostic methods are now widely available in well-equipped medical structures, there is still need for the development of simpler, smaller, easy to use and lower cost diagnostic tools. Although less prominent, the same requirements apply for analytical-grade chemical sensing systems in environmental monitoring, safety and security.

Continuous developments in MEMS technology and microfluidics are key drivers for the miniaturization of lab-grade sensing systems. Micro-technologies and miniaturization allow for designing lightweight and small devices, but other advantages relevant though less obvious: reduced consumption of power and reagents, faster response times, increased sensitivity, reduced environmental footprint, availability of batch production processes for low-cost and disposable devices.

This special issue publishes 7 novel contributions in the fields of biosensing, lab-on-chip, organ-on-chip and related technologies such as numerical microfluidics studies, digital micro-fluidics and micromixers.

In [1], Yuan Tian et al. report on a novel MEMS microcantilever biosensor fabrication process which does not rely on expensive and potentially fragile SOI wafers. The novel design uses titanium as piezo-resistor between two thin Polyimide and SiO_2 passive layers, featuring a comparatively low spring constant. The cantilevers are used in a Wheatstone bridge configuration and show a very low noise, and sensing experiments with functionalised cantilevers were performed. A minimum detectable ricin concentration of 10 ng/mL is reported, with a good linear behaviour over the 0–80 ng/mL concentration range.

The extremely low abundance of biomolecules in most sensing applications is the most well-known challenge. With the aim of reaching ever lower bio-detection limits, Yan Chen et al. [2] report on the development of an acoustic concentrator, named "acoustic tweezers", which uses high frequency longitudinal acoustic waves to promote the concentration of the target molecules in the detection area of an electrolyte-gated graphene field effect transistor. The effect of the acoustic tweezers was validated under the microscope by IgG labelled with green fluorescence, and by comparison of the FET response (Dirac point shift) with and without acoustic accumulation.

To reach the goal of developing lower-cost and easy to use point-of-care analysis systems, Qin Huang et al. [3] propose a novel microfluidic lab-on-chip for isothermal parallel detection of multiple targets. The approach is based on a two-stage recombinase polymerase amplification and subsequent loop-mediated isothermal amplification (RPA + LAMP) microfluidic platform, which is demonstrated in a disc-shaped milled PMMA lab-on-chip. The system is able to detect different types of nucleic acid targets, e.g., bacteria and RNA viruses, at the same time. The detection is comparatively fast and

Citation: Zampolli, S. Editorial for the Special Issue on MEMS and Microfluidic Devices for Analytical Chemistry and Biosensing. *Micromachines* **2022**, *13*, 896. https://doi.org/10.3390/mi13060896

Received: 1 June 2022
Accepted: 2 June 2022
Published: 4 June 2022

operates at two isothermal setpoints, with high sensitivity and specificity. Disc rotation at up to 2000 rpm is used for priming and sample transfer.

Digital microfluidics (DMF) are a very efficient technology to dispense, move, mix and transport micro-droplets of sample, reagents and reaction products. Electro-wetting on dielectric (EWOD) DMF are proposed by Zhen Gu et al. [4] for colorimetric sensing based on gold nanoparticles. The authors use a PCB-based DMF chip with a PTFE film as dielectric and hydrophobic layer and 0.1 mm electrode gaps. The contact angle of the gold nanoparticle solution was compared with pure water, resulting in a higher voltage required to move the suspension of nanoparticles. Evaporation effects were studied by UV/Vis absorbance measurements, and no significant evaporation was found after a 0.6 m travel distance. Colorimetric sensing tests were also performed targeting Hg^{2+}, using 2 LED sources and 2 light detectors, and demonstrated a detection limit of 0.01 μMol.

In addition to the biosensors reported above, another increasingly prominent application of microfluidics is in the area of organ-on-chip devices. Larry J. Millet et al. [5] report on a platform to study radiation damage to pulmonary lumens, with the specific target of identifying candidate biomarkers of radiation damage. In this very interesting and novel pilot study a PDMS microfluidic platform is fabricated by molding from a Si master and loaded with HMVEC-L cell cultures to assess radiation damage after exposure to 10 Gy of Co-60 radiation. The original design of the first device was optimized by adding a third cell seeding port, to increase the cell loading uniformity. The perfusates of the cell cultures exposed to gamma irradiation were studied by LC-MS/MS and data searched against human proteome database, finding 26 statistically significant proteins that change in abundance between irradiated and nonirradiated tissue chip platforms.

While MEMS and microfluidics enable the miniaturization of the core sensing systems, in-field deployment often requires the miniaturization of auxiliary devices as well. In analytical chemical gas sensing applications, periodic recalibration is necessary, and in some cases the very low flow requirements and the high sensitivity of the sensors require novel approaches for an efficient calibration gas generation. Florian Noël et al. [6] present a low concentration gas mixture generator for the in-field calibration of a BTEX GC detector, which solves the problem of unnecessary consumption of both dilution and calibrated gases. The reported device is the experimental validation of a design studied by CFD in a previous paper. It uses a time-pulse approach and a micro-mixer device milled into PMMA, reaching dilutions up to 1:1000 with good precision and reproducibility at a low total gas flow of 25 NmL/min.

Phenomena at the microscale are sometimes difficult to observe, and novel techniques to characterize flow velocities in microchannels include the photobleaching effect. In an in-depth numerical study, Yu Chen et al. [7] use a model based on the convection–diffusion reaction equation to study the photobleaching process with reference to a gaussian laser focus region. The study allowed for the determination of the profiles of effective dye concentration and fluorescence, and to find the relationship between the photobleaching time constant obtained by experiments and the photochemical reaction coefficient.

I want to thank all the authors and teams who have contributed to the papers for this first volume of the Micromachines Special Issue on MEMS and Microfluidic Devices for Analytical Chemistry and Biosensing.

A special thank goes also to the academic editors Katsuo Kurabayashi and Aiqun Liu for their precious contribution, and to Jerry Chen and Min Su of the Micromachines Editorial Office for their great support.

Funding: This research received no external funding.

Conflicts of Interest: The author declares no conflict of interest.

References

1. Tian, Y.; Zhao, R.; Liu, Y.; Yu, X. A Low Spring Constant Piezoresistive Microcantilever for Biological Reagent Detection. *Micromachines* **2020**, *11*, 1001. [CrossRef] [PubMed]
2. Chen, Y.; Liu, W.; Zhang, H.; Zhang, D.; Guo, X. A Sensitivity-Enhanced Electrolyte-Gated Graphene Field-Effect Transistor Biosensor by Acoustic Tweezers. *Micromachines* **2021**, *12*, 1238. [CrossRef] [PubMed]
3. Huang, Q.; Shan, X.; Cao, R.; Jin, X.; Lin, X.; He, Q.; Zhu, Y.; Fu, R.; Du, W.; Lv, W.; et al. Microfluidic Chip with Two-Stage Isothermal Amplification Method for Highly Sensitive Parallel Detection of SARS-CoV-2 and Measles Virus. *Micromachines* **2021**, *12*, 1582. [CrossRef] [PubMed]
4. Gu, Z.; Luo, J.-J.; Ding, L.-W.; Yan, B.-Y.; Zhou, J.-L.; Wang, J.-G.; Wang, H.-F.; Kong, C. Colorimetric Sensing with Gold Nanoparticles on Electrowetting-Based Digital Microfluidics. *Micromachines* **2021**, *12*, 1423. [CrossRef] [PubMed]
5. Millet, L.; Giannone, R.; Greenwood, M.; Foster, C.; O'Neil, K.; Braatz, A.; Davern, S. Identifying Candidate Biomarkers of Ionizing Radiation in Human Pulmonary Microvascular Lumens Using Microfluidics—A Pilot Study. *Micromachines* **2021**, *12*, 904. [CrossRef] [PubMed]
6. Noël, F.; Trocquet, C.; Serra, C.; Le Calvé, S. Experimental Validation of a Novel Generator of Gas Mixtures Based on Axial Gas Pulses Coupled to a Micromixer. *Micromachines* **2021**, *12*, 715. [CrossRef] [PubMed]
7. Chen, Y.; Meng, S.; Wang, K.; Bai, J.; Zhao, W. Numerical Simulation of the Photobleaching Process in Laser-Induced Fluorescence Photobleaching Anemometer. *Micromachines* **2021**, *12*, 1592. [CrossRef] [PubMed]

micromachines

MDPI

Article

A Low Spring Constant Piezoresistive Microcantilever for Biological Reagent Detection

Yuan Tian [1], Rui Zhao [2], Yi Liu [1] and Xiaomei Yu [1,*]

1 National Key Laboratory of Science and Technology on Micro/Nano Fabrication, Institute of Microelectronics, Peking University, Beijing 100871, China; nisioisin@pku.edu.cn (Y.T.); yiliu1@pku.edu.cn (Y.L.)
2 Science and Technology on Electronic Test and Measurement Laboratory, North University of China, Taiyuan 038507, China; zhaorui@nuc.edu.cn
* Correspondence: yuxm@pku.edu.cn

Received: 22 October 2020; Accepted: 11 November 2020; Published: 12 November 2020

Abstract: This paper introduces a piezoresistive microcantilever with a low spring constant. The microcantilever was fabricated with titanium (Ti) as the piezoresistor, a low spring constant polyimide (PI) layer, and a thin silicon oxide (SiO_2) layer as the top and bottom passive layers, respectively. Excellent mechanical performances with the spring constant of 0.02128 N/m and the deflection sensitivity $(\Delta V/V)/\Delta z$ of 1.03×10^{-7} nm^{-1} were obtained. The output voltage fluctuation of a Wheatstone bridge, which consists of four piezoresistive microcantilevers, is less than 3 μV@3 V in a phosphate buffered saline (PBS) environment. A microcantilever aptasensor was then developed through functionalizing the microcantilevers with a ricin aptamer probe, and detections on ricin with concentrations of 10, 20, 50 and 100 ng/mL were successfully realized. A good specificity was also confirmed by using bovine serum albumin (BSA) as a blank control. The experiment results show that the Ti and PI-based microcantilever has great prospects for ultrasensitive biochemical molecule detections with high reliability and specificity.

Keywords: polyimide; piezoresistive microcantilever; spring constant; deflection sensitivity; ricin

1. Introduction

The microcantilever was developed some time ago as a miniature biochemical sensor with high sensitivity. Its readout methods can be classified as optical method [1,2], piezoresistive method [3–5], and capacitive method [6,7]. Although traditional optic analytical methods have relatively high sensitivity, they require large, expensive laser instruments and well professional and technical personnel to operate them. Compared to optical analytical methods, a piezoresistive microcantilver-based sensor is easy to use by reading out the signals with high precision multimeter or complementary metal oxide semiconductor (CMOS) circuits. It can also be applied in a liquid environment without sensitivity loss [8].

A single crystalline silicon-based microcantilever fabricated by silicon-on-insulator (SOI) wafer is widely used to realize biochemical detections. In 2017. Zhao el al. reported a Si-based piezoresistive microcantilever that was able to detect Dimethyl methylphosphonate (DMMP) with a sensitivity of 1.0 μM, and the spring constant of the microcantilver was approximately 0.04679 N/m [9]. In 2017, Patkar et al. created a dynamic mode piezoresistive microcantilever with a spring constant of 0.2 N/m, a deflection sensitivity of 0.3 ppm nm^{-1}, and a resonant frequency at 22.5 kHz [10]. In 2018, Ku et al. invented a piezoresistive microcantilever using SiO_2, Si_3N_4, and polysilicon, which realized a minimum detection limit of C-reactive protein (CPR) concentration at 100 μg with 3.1 N/m surface stress change by a temperature compensation [11]. Xu et al. introduced a piezoresistive microcantilever with silicon nanopillars and ZnO nanorod on the top of the surface, which achieved an increase of the surface area by a factor of about 100 and a 2.1 ppb detection limit of NO_2 detection [12,13]. In 2020, Rotake et al.

studied a piezoresistive microcantilever for Hg^{2+} detection with a limit of 0.75 ng/mL. The spring constant of the microcantilever was 0.07053 N/m [14]. In 2020, Kandpal et al. reported a piezoresistive microcantilever which realized a 1,3,5-trinitroperhydro-1,3,5-triazine (RDX) detection with 6-MNA surface functionalization. The spring constant was about 0.2 N/m and the deflection sensitivity was approximately 0.161 ppm/nm. RDX vapors were detected within about 30 s exposure time [15].

To fabricate a microcantilever with higher sensitivity, a thin single crystalline silicon layer is required, which can only be obtained from an expensive silicon-on-insulator (SOI) wafer, which is easily broken during applications because of the high spring constant. Flexible piezoresistive microcantilevers were developed with metal as the piezoresistor and a polymer as the passive layer in recent years. In 2011, Zhu et al. studied a V-shaped piezoresistive microcantilever with polyimide as the structural layer. The polyimide (PI)-based microcantilever has 17.5 μm length and 1.45 μm thickness. The deflection sensitivity of the microcantilever is $1.1 \times 10^{-6}/\text{nm}^{-1}$, and the detection limit of TNB gas is 10 ppb [16]. In 2013, Liu et al. reported a rectangular microcantilever with composite polyimide as the structural layer and 40 nm gold as adhesion layer. The length, width, and thickness of the microcantilever are 1000 μm, 200 μm, and 25 μm, respectively, which achieved the detection of surface stress change of 5.66 N/m caused by YN94 growth [17]. These microcantilevers are relatively thick and the sensitivity is very limited.

In this work, a flexible microcantilever was fabricated on a silicon wafer with Titanium (Ti) as the piezoresistor and thin polyimide (PI) and SiO_2 as the passive layers. A spring constant of 0.02128 N/m and a deflection sensitivity of 1.03×10^{-7} nm^{-1} were successfully achieved.

2. Design and Fabrication

2.1. Design of Piezoresistive Microcantilever

Polyimide is widely used in MEMS device for its low residual stress, low coefficient of thermal expansion, low water absorption, and good insulation performance. PI 2610 has a Young's modulus of 8.5 Gpa. To obtain a low spring constant microcantilever, 1.3 μm thick PI was selected as the top passive layer and a 100 nm thick SiO_2 as the bottom layer in this work.

For a piezoresistive microcantilever, the sensitivity is defined as the relative resistance change ($\Delta R/R$) versus the free end vertical distance change (Δz), which can be described as [18]:

$$\frac{\Delta R/R}{\Delta z} = \frac{3Kt(l_c - l_{leg}/2)}{2l_c^3} \tag{1}$$

where K is the gauge factor, t is the thickness, l_c is the length of the microcantilever, and l_{leg} is the piezoresistor length. The gauge factor is defined as [19]:

$$K = (1 + 2\nu) + \frac{\Delta\rho/\rho}{\varepsilon} \tag{2}$$

where ν is Poisson's ratio of the piezoresistor, ε is the length change of the piezoresistor, $\Delta\rho$ is the electrical resistivity change of the piezoresistor, and ρ is the original electrical resistivity of the piezoresistor. To increase the piezoresistive microcantilever's sensitivity, we must increase the gauge factor and thickness of the microcantilever, and decrease its length. Although single crystalline silicon has a high piezoresistive coefficient, its spring constant and the fabrication cost are higher, since the thin single crystalline silicon layer can only be obtained from an expensive silicon-on-insulator (SOI) wafer. Another type of piezoresistive material is metal. For metal materials, their electrical resistivity is quite stable at the same temperature. $\Delta\rho/\rho$ has almost no contribution for $\Delta R/R$. The most contributing parts of the resistance change are Poisson's ratio and the length change. Most metal materials show a relatively high positive electrical temperature coefficient of resistance (TCR), so a low TCR piezoresistive material is expected to improve the microcantilever stability. Titanium has a minimum TCR of 0.00063 °C^{-1} and medium Poisson's ratio of 0.34 [20,21] among all metal materials, hence, it was implemented as the piezoresistive material to ensure stable output voltage and high sensitivity in our design.

The sensitivity of the microcantilever is also closely related to its length and thickness, therefore, it is necessary to carefully optimize the parameters of the microcantilever for it to be able to detect biochemical molecules. To avoid the microcantilever sticking to the substrate and keep a low spring constant at the same time, a rectangular PI/Ti/SiO_2 piezoresistive microcantilever was designed with a dimension of 100 μm in length, 50 μm in width, and 1.44 μm in thickness. To ensure high sensitivity, the piezoresistor was placed at the root of the microcantilver. Also, the length should be as long as possible to decrease thermal drift at the same time. Thus, a four-fold shape piezoresistor with a single leg dimension of 60 μm × 5 μm × 40 nm was designed. With the above designed parameters, the sensitivity of the microcantilever is calculated to be 3.645×10^{-7} nm^{-1} according to Equation (1).

As shown in Figure 1, the microcantilever consists of a 1.3 μm thick PI top passive layer, a 40 nm thick titanium piezoresistive layer, and a 100 nm thick SiO_2 bottom passive layer from top to bottom. To reduce the impact of environmental noise and initial thermal mismatch, four microcantilevers form a sensor set of Wheatstone bridge. The Wheatstone bridge was configured as half-bridge type as shown in Figure 1 [22], in which nonadjacent two were coated with a 10/40 nm Cr/Au modified layer as the sensing microcantilevers, which will bent and the resistance of the piezoresistor will be changed due to the reaction process between the sensing and target molecules, while other two are only with PI on the top as reference microcantilevers. The differential signal output of the Wheatstone bridge V_{out} can be described as:

$$V_{out} = \frac{1}{2} V_{in} \frac{\Delta R}{R} \tag{3}$$

where V_{in} is the supplied voltage, R is the original resistance of the piezoresistor, and ΔR is the resistance change of the piezoresistor. According to Equations (1) and (3), the sensitivity of the microcantilever can be finally written as:

$$\frac{\Delta V/V}{\Delta z} = \frac{3Kt(l_c - l_{leg}/2)}{4l_c^3} \tag{4}$$

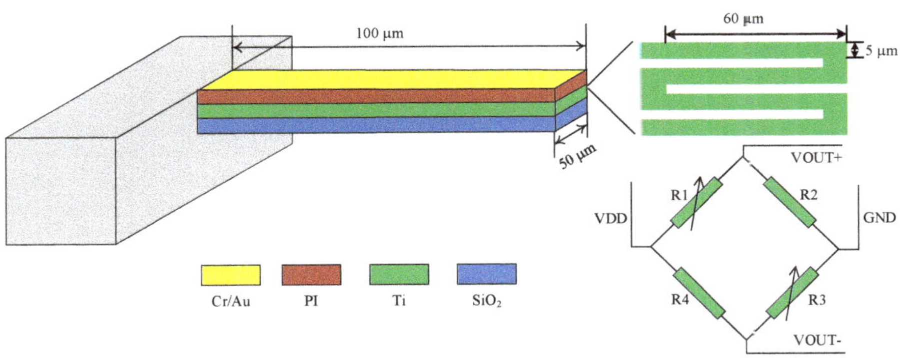

Figure 1. The schematic diagram of the polyimide (PI)/titanium (Ti)/silicon oxide (SiO_2) microcantilever. The shape of the piezoresistor is four-fold and four microcantilevers consist of a Wheatstone bridge.

2.2. Fabrication

The microcantilever was fabricated on a 525 μm thick N-type (100) single-polished silicon wafer. As shown in Figure 2, the entire fabrication process includes following a 5-step photolithography process:

(a) Firstly, LPCVD technology was used to deposit a 100 nm thick SiO_2 on the surface of the silicon wafer as the bottom passive layer of the microcantilever. After the first photolithography, BHF was used to remove the SiO_2 outside the pattern to define the SiO_2 layer of microcantilever.

(b) Then, a 40 nm thick Titanium was sputtered and patterned to form the Ti piezoresistor.
(c) An 800 nm thick Al was sputtered, and Al connection wires and pads were patterned in the third photolithography.
(d) After that, a 1.3 μm thick PI was spin coated on the surface of the substrate at 3000 rpm for 30 s. The wafer was then annealed in a nitrogen atmosphere with a gas flow of 4.5 L/min and a temperature of 350 °C for 90 min to complete the PI layer curing. Apart from the piezoresistor, the PI layer also fully encapsulates the aluminum wires to avoid electrical conduction in an aqueous environment.
(e) E-beam evaporation was used to deposit a 10/50 nm Cr/Au modified layer on the surface of the PI layer. Cr was used as the adhesion layer between PI and Au. After the fourth photolithography, the Au/Cr was etched by wet etching technology to form a biochemical molecule modified layer in sensing microcantilevers.
(f) At last, the microcantilever was patterned and defined by using a 6 μm thick photoresist as a mask. After PI was etched by anisotropic RIE oxygen plasma etching until the silicon wafer, a dry silicon etching techniques was used to etch the silicon wafer from front side until the microcantilever structure was completely released, and a reactive well was also formed at the same time. The traditional release of the microcantilever structure by the isotropic dry etching technology can cause serious lateral undercutting situation problems at the sidewall near the fixed end of the microcantilever. A combination of anisotropic and isotropic dry etching techniques was used to avoid this lateral undercutting problem of SiO_2.

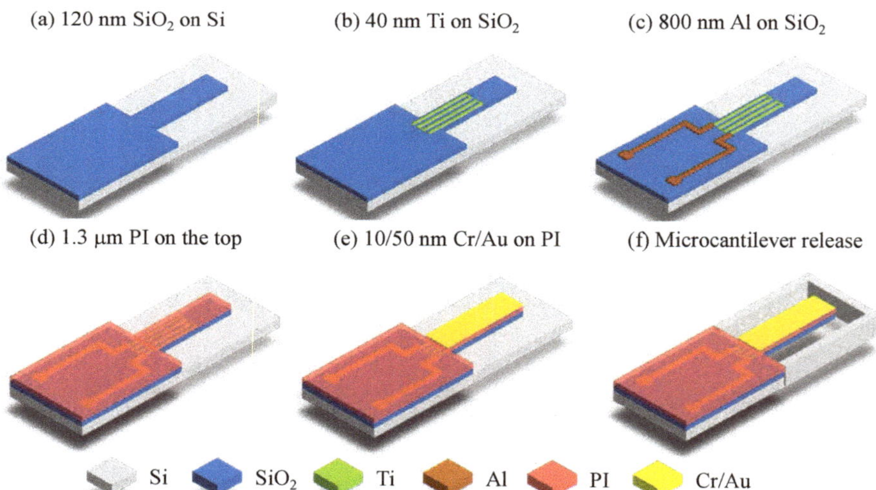

Figure 2. The fabrication processes of PI/Ti/SiO2 microcantilevers. (**a**) LPCVD technology was used to deposit 100 nm SiO_2. (**b**) Sputtering was used to make 40 nm thick Titanium piezoresistor. (**c**) Sputtering was used to deposit 800 nm thick Al pad. (**d**) Spin coating was implemented to make 1.3 μm thick PI. (**e**) Sputtering was used to deposit 10/50 nmthick Cr/Au modified layer. (**f**) A combination of anisotropic and isotropic dry etching techniques was used to etch the silicon wafer until the microcantilever structure is completely released.

A laser confocal microscope OLS 5000 was introduced to rebuild a 3D vision of the microcantilever, as shown in Figure 3 Due to the optic projection, the hollow space below the microcantilever cannot be well displayed in Figure 3a. As shown in Figure 3b, a deep reactive well fabricated by the anisotropic dry etching can be seen clearly. Figure 3c is the height scope of the microcantilever; it can be seen that the microcantilever bent upward for 36.122 µm and the depth of the reactive well is 40.534 µm. Since the coefficient of thermal expansion for PI is higher than that of Si in this temperature range, the tensile stresses in the PI films [14] made the microcantilever bend upward, which can inhibit the microcantilever from sticking to the bottom of the reactive well.

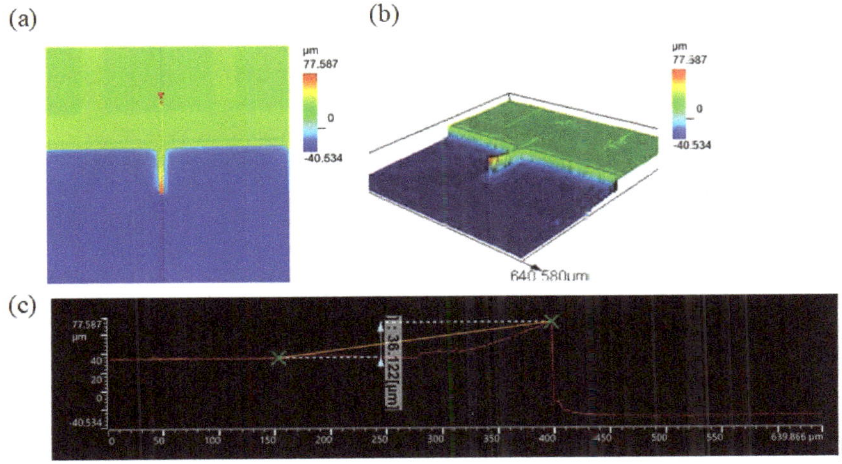

Figure 3. 3D measurement and observations images by laser confocal scanning microscope LEXT OLS5000 (**a**,**b**) are the top and 3D visions of a microcantilever. The height of free end is 77.587 µm. (**c**) is the height scope of the microcantilever. The upper bend of the free end of the microcantilever is 36.122 µm and the depth of reactive well is 40.534 µm.

A full vision of the microcantilever was also introduced by scanning electronic microscopy (SEM) to get a higher quality structure image. Figure 4a is the SEM image of a Wheatstone bridge with four microcantilevers, and Figure 4b is a sensing microcantilever.

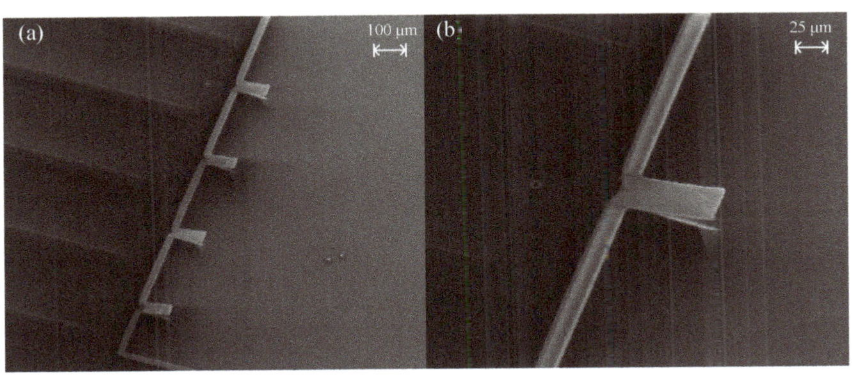

Figure 4. SEM image of (**a**) four microcantilevers that make up the Wheatstone bridge and (**b**) a sensing microcantilever respectively.

3. Performances

3.1. Spring Constant

It is difficult to calculate the spring constant k for a multi-layers microcantilver, so we measured it with a reference cantilever method. An AFM probe (MFP-3D-BIO PNP-DB) manufactured by Nanoworld in Switzerland was used as a reference microcantilever. The AFM probe has a rectangular structure with a geometric size of 200 μm × 13 μm × 0.6 μm, an intrinsic resonance frequency of 17 kHz, and a spring constant of 0.06 N/m. Figure 5a shows the measured relationship between the deflection of the microcantilever and the displacement of the piezoelectric driven AFM probe under a load. The red curve in the figure represents the approach process of the AFM probe, and the blue curve represents the AFM probe retract process. As the AFM tip approaches the microcantilever surface, initially the forces are too small to give a measurable deflection of the AFM tip. Then the attractive forces (such as Van der Waals) overcome the AFM tip spring constant and the tip jumps into contact with the microcantilever surface. Once the AFM tip contact the microcantilever's surface, it remains on the surface as the separation between the AFM tip and microcantilever surface, causing a deflection of the AFM tip and the microcantilever. As the AFM tip is retracted from the microcantilever surface, often the AFM tip remains in contact with the surface due to some adhesion and the AFM tip is deflected downwards. At some distance, the force from the AFM tip can finally overcome the adhesion and get to the initial stage [23]. Based on the AFM probe approach process and spring constant of AFM probe, we can establish the relationship between the applied force and the deflection displacement of the microcantilever as shown in the Figure 5b. The spring constant of the microcantilever is obtained to be 0.02128 N/m by the fitted line slope, which is almost half of the one compared with the silicon-based microcantilever of 0.04679 N/m [24]. The adjustment coefficient of the fitting result is adj. R^2 = 0.998, indicating that the microcantilever has a good linear response for applications of the PI/Ti/SiO_2 microcantilever

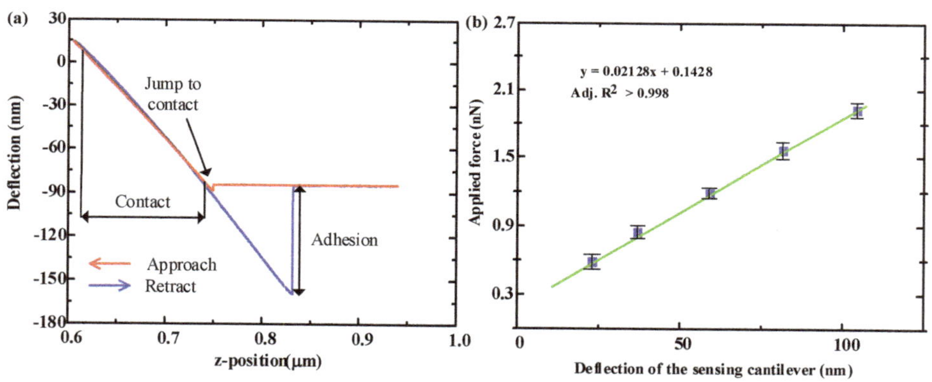

Figure 5. (**a**) The relationship curve between the deflection displacement of the microcantilever and the AFM tip position change under the AFM load force. (**b**) The linear correlation of the relationship between the load force on a microcantilever and its deflection displacement.

3.2. Sensitivity

The deflection sensitivity is $(\Delta V/V)/\Delta z$, where Δz is vertical displacement change of free end of the microcantilever. In order to obtain the deflection sensitivity, the microcantilever was fixed on a probe station in which the probe was driven by a precision electronically controlled translation stage with a minimum step size of 5 μm. During the test, the probe put a pressure at the free end of the microcantilever, and thus changed the piezoresistor value. The output voltage of the microcantilever was measured by an Agilent 34,401 A $6^{1/2}$ mustimeters. The relative voltage change ($\Delta V/V$) versus the vertical displacement change Δz for the $PI/Ti/SiO_2$ piezoresistive microcantilever is shown in Figure 6. By linearly fitting the data between the output voltage change of the microcantilever and the vertical displacement of the microcantilever, a deflection sensitivity of 1.03×10^{-7} nm^{-1} was obtained. The determination coefficient of the fitting straight line Adj. R^2 is above 0.99, indicating that the flexible $PI/Ti/SiO_2$ microcantilever has a good linear relationship.

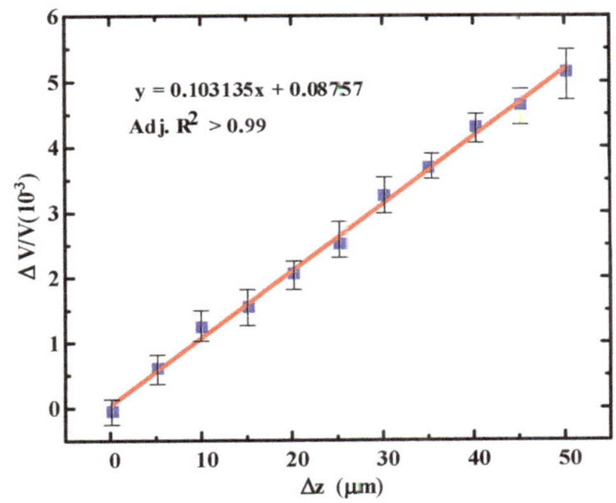

Figure 6. The relationship between the vertical displacement change Δz of the free end of the microcantilever and the relative voltage change $\Delta V/V$.

3.3. Stability

The biochemical detection experiments involved in this work were carried out in a liquid phase environment containing inorganic salts. In order to verify the feasibility of the microcantilever for trace detections, we tested the stability of the sensor output signal in phosphate buffered saline (PBS) with a pH of 7.4 and a concentration of 0.01 M. Firstly, the microcantilever chip was fixed on a printed circuit board (PCB) and electrically connected to the PCB through wire bonding. Then, a 3 V DC bias voltage was applied to the bridge, and the output differential voltage signal was measured by an Agilent 34,401 A $6\frac{1}{2}$ high-precision multimeter, and real-time data acquisition was performed by the upper computer. Figure 7 shows the test results of the stability of the output signal of the microcantilever in the PBS solution. It can be observed from the figure that the output signal of the sensor increases rapidly during the initial period of time (about 20 min), which is caused by the self-heating of the titanium piezoresistor under the bias voltage. Due to the different thermal expansion coefficients of PI and SiO_2, the microcantilevers have different thermal deformations, which ultimately lead to the output signal changing in the initial stage. Subsequently, the microcantilever output signal reached a state of dynamic equilibrium. The temperature cross sensitivity played a minor role during the detection, while the 1/f noise and Johnson noise of the piezoresistors played a leading role. The fluctuation range of its output voltage was less than 3 μV as shown in the inset of Figure 7, and this experimental results show that the microcantilever prepared in this paper has low output voltage noise in PBS solution, and can meet the demand for the trace level detections of biochemical molecules.

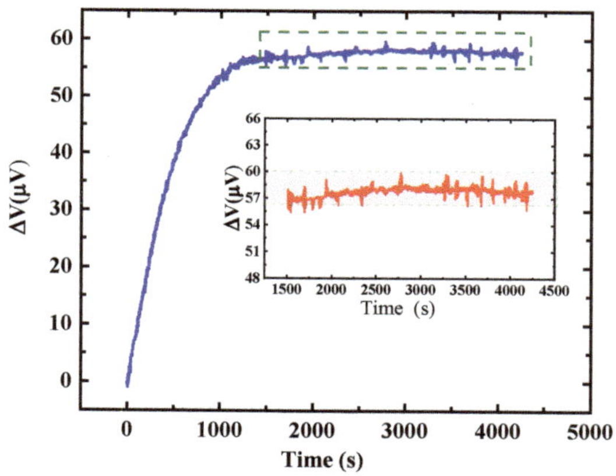

Figure 7. The output voltage fluctuation of a microcantilever in PBS buffer (PH 7.4, 0.01 m) at temperature (20 °C) and 3 V bias voltage. Inset shows the voltage fluctuation is less than 3 μV.

4. Biological Detections

4.1. IgG Detections

As a regular biosensor detection indicator [25–28], IgG was first introduced to evaluate the detection sensitivity of the microcantilever. Related experiment materials of 3,3′-Dithiopropionic Acid (DDPA), 1-Ethyl-3-(3-dimethylaminopropyl) Carbodiimide Hydrochloride (EDC), N-Hydroxy Succinimide (NHS), Streptavidin, Ethanolamine, bovine serum albumin (BSA), and PBS solutions were all purchased from Sigma-Aldrich (Shanghai, China). Biotinylated goat anti-human IgG (H+L) polyclonal antibody was purchased from Bioss Antibodies Inc. (Beijing, China). Concentrated sulfuric acid (H_2SO_4, ≥95–98%), hydrogen peroxide (H_2O_2, ≥30%), and concentrated hydrochloric acid (HCl, ≥38%) were purchased from Beijing Chemical Works (Beijing, China).

Before the biological sensing detection experiment, the microcantilever needed to be pre-processed to remove organic contaminants on the surface of the microcantilever. The pretreatment process of the microcantilever was as follows: First, the microcantilever biosensor was treated with RIE oxygen plasma for 30 s in RF power of 250 W, oxygen flow rate of 20 sccm. Then, the microcantilever biosensor was placed in acetone for 20 min, and this was repeated twice. Subsequently, the microcantilever biosensor was placed in absolute ethanol for 20 min. Finally, the microcantilever biosensor was cleaned with deionized water several times and left to dry naturally in a nitrogen atmosphere.

For detecting IgG, the microcantilever sensors were constructed by immobilizing a biotinylated goat anti-human IgG on the surface of a sensing microcantilever with antibody concentration of 100 μg/mL by using the biotin streptavidin system (BAS) method. The sensing microcantilever surface functionalization process was as follows:

(a) The pre-processed microcantilever was immersed in 5 mg/mL DDPA for about 1 h to coat the carboxyl groups on the Au film surface of the sensing microcantilever
(b) Then, it was soaked in the EDC/NHS mixture with a concentration of 5 mg/mLs, and a volume ratio of 3:1 for about 30 min to form a succinimide activator on the surface of the sensitive microcantilever.
(c) After being repeatedly cleaned, the microcantilever was immersed in a 0.1 mg/mL streptavidin solution to react for about 1 h to cross-link the streptavidin molecule with the succinimide activator;
(d) At last, a 1 M ethanolamine was injected to inactivate the residual carboxyl groups on the surface of the sensing microcantilever for 30 min after cleaning the microcantilever a few times.

After the functionalization, human IgG with concentrations of 10 ng/mL, 20 ng/mL, 30 ng/mL, 50 ng/mL, and 80 ng/mL were detected with the sensors, the test results are shown in Figure 8a. The relationship between the microcantilever response voltage and IgG concentration is shown in Figure 8b. Within the detection concentration range of IgG, the biosensor showed a good linear response, and its detection limit, the minimum detectable concentration, reached 10 ng/mL. The response time of the sensor is the time when the output voltage signal changed from the injection moment to stable saturation state. The relationship between the microcantilever response time and the IgG concentration is shown in Figure 8c. The response time from injection to dynamic equilibrium were 1835 s, 2246 s, 3120 s, 3654 s, and 4260 s for the concentrations of IgG 10 ng/mL, 20 ng/mL, 30 ng/mL, 50 ng/mL and 80 ng/mL respectively. The limiting factor for response time depends on a complicated dynamic process of immobilized biotinylated goat antihuman IgG and IgG. As the concentration of target IgG increased, the output voltage signal of the sensor becomes larger, and the response time becomes longer. At the same time, the specificity was also tested with 100 ng/mL BSA as a blank control, and the IgG biosensor basically did not respond to it.

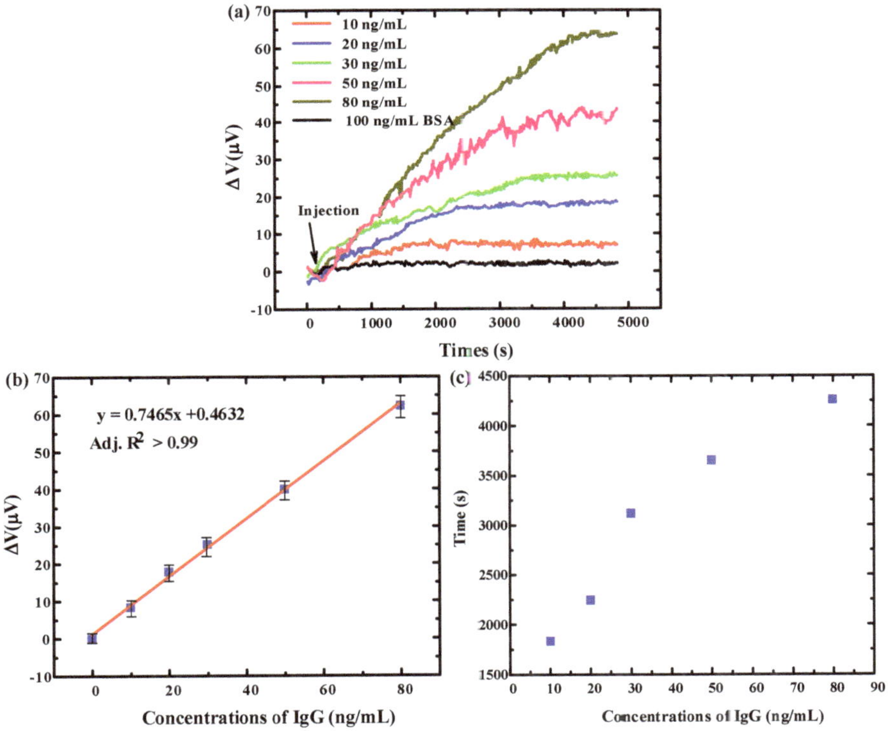

Figure 8. (**a**) The response output curve of PI/Ti/SiO_2 flexible microcantilever microcantilever to IgG detection (**b**) The linear relationship between the microcantilever response voltage and IgG concentration. (**c**) The relationship between the microcantilever response time and the IgG concentration.

4.2. Ricin Detection

Ricin is a toxic protein extracted from castor beans. Several traditional analytical techniques for detecting ricin are based on the optical method such as mass spectrometry [29], Raman spectrometry [30] and fluoroimmunoassay [31,32]. A microcantilever aptamer sensor for detecting ricin was constructed by immobilizing a ricin probe on the surface of sensing microcantilever. The DNA sequence of the ricin aptamer is 5′-SH-TCG CAA GAC GGA CAG AAG CTG ATT GTTATT TTT TTT TTT GTT TAT GCT GTA TGC CAT TAG GTT GGT GGA GCG ATT TGT-3′ and was synthesized by solid-phase peptide synthesis and purified until it reached 95% purity by polyacrylamide gel electrophoresis (PAGE) method (AuGCT. LTD., Beijing, China). The thiol modified ricin aptamer was stored at −18 °C for the ricin experiments. Other reagents and materials are the same as mentioned in Section 4.1.

The microcantilever biosensor was cleaned before the functionalization process to remove organic contaminants on the surface of the microcantilever. The cleaning process of the microcantilever biosensor was as follows: First, the microcantilever biosensor was treated with RIE oxygen plasma for 30 s in RF power of 250 W, oxygen flow rate of 20 sccm. Then, the microcantilever biosensor was placed in acetone for 20 min, repeated for twice. Subsequently, the microcantilever biosensor was placed in absolute ethanol for 20 min. Finally, the microcantilever biosensor was cleaned with deionized water several times and dried naturally in a nitrogen atmosphere.

During the fabrication process, the microcantilever was immersed in a sulfhydryl aptamer solution with a concentration of 20 μM, and incubated for 8 h at room temperature to coat the sulfhydryl aptamer on the surface of the sensitive microcantilever to complete the probe immobilization. After washing the microcantilever sensor chip with PBS solution several times to elute the unfixed and weakly bound aptamer probe, 1% BSA was used to seal the surface of the sensitive beam for about 30 min. Finally, the chip was cleaned several times to elute the unfixed BSA, thus completing the functionalization process of the microcantilever.

For the detections, the ricin solution was diluted to 0.2 μg/mL, 0.4 μg/mL, 0.8 μg/mL, 1.2 μg/mL and 2 μg/mL solutions by using the PBS solution beforehand. Firstly, the microcantilever was placed in a reaction well containing 400 μL PBS solution and biased a 3 V voltage until the sensor output signal achieved a stable state. Then, the diluted ricin solutions were injected to the reaction well and to form detection concentrations of 10 ng/mL, 20 ng/mL, 40 ng/mL, 60 ng/mL and 100 ng/mL respectively. The output signals were measured by an Agilent 34,401 A $6\frac{1}{2}$ high-precision multimeter with a sample frequency of 0.5 Hz. Figure 9a shows the sensor's detection results of ricin at concentrations from 10 ng/mL to 100 ng/mL. The dynamic equilibrium voltage variations were 9 μV, 15 μV, 25 μV, 37 μV, and 55 μV for the concentrations of ricin 10 ng/mL, 20 ng/mL, 40 ng/mL, 60 ng/mL and 100 ng/mL respectively. Figure 9b shows the relationship of microcantilever response voltage and detection concentration; the output voltage of the sensor increases linearly with the concentration of ricin, with the minimum detectable concentration of 10 ng/mL. The relationship between the microcantilever response time and the ricin concentration is shown in Figure 9c. The response time from injection to dynamic equilibrium were 2645 s, 3326 s, 4121 s, 4452 3, and 4875 s for the concentrations of ricin 10 ng/mL, 20 ng/mL, 40 ng/mL, 60 ng/mL and 100 ng/mL, respectively. The limiting factor for response time depend on a complicated dynamic process of the ricin aptamer and ricin [33]. At the same time, abrin, which is also a dimeric glycoprotein, was used as the target molecule to evaluate the specificity of the ricin aptamer sensor. The sensor did not show a significant response to 100 ng/mL abrin, which reflected a good specificity.

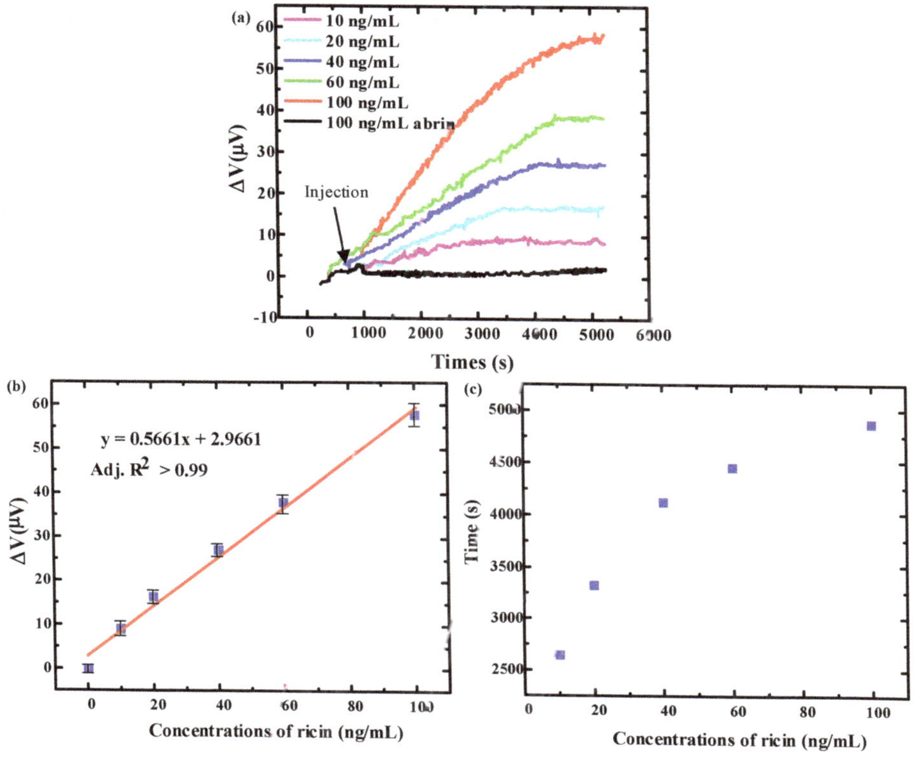

Figure 9. (**a**) The output voltage curve of the microcantilever aptamer sensors at detecting concentration of 10–100 ng/mL ricin (**b**) Linear relationship between microcantilever response voltage and detection concentration. (**c**) The relationship between the microcantilever response time and the ricin concentration.

5. Conclusions

In this work, a low spring constant piezoresistive microcantilever was designed and fabricated using a Ti film as the piezoresistor and a low spring constant PI and a thin SiO_2 as the top and the bottom insulation layers, respectively. The microcantilever has excellent mechanical properties with 0.02128 N/m spring constant and 1.03×10^{-7} nm^{-1} deflection sensitivity. The microcantilever also shows a good electrical stability with less than 3 μV voltage fluctuation in PBS buffer. After surface functionalization with thiol modified ricin aptamer, the microcantilever was introduced to detect ricin with 10 ng/mL minimum detectable concentration and good linear correlation results. All experiment results show that the flexible PI/Ti/SiO_2 microcantilever prepared in this paper has good biological detection performance and can be used for highly sensitive biomolecule detection.

Author Contributions: Methodology, R.Z., X.Y.; validation, Y.T. and X.Y.; formal analysis, Y.T.; investigation, R.Z.; resources R.Z.; data curation, R.Z., Y.T. and Y.L.; writing—original draft preparation, Y.T.; writing—review and editing, X.Y.; visualization, Y.T.; supervision, X.Y.; project administration, X.Y.; funding acquisition, X.Y. All authors have read and agreed to the published version of the manuscript.

Funding: The research was funded by the National Natural Science Foundation of China (grants 61935001).

Acknowledgments: In this section you can acknowledge any support given which is not covered by the author contribution or funding sections. This may include administrative and technical support, or donations in kind (e.g., materials used for experiments).

Conflicts of Interest: The authors declare no conflict of interest.

References

1. Meyer, G.; Amer, N.M. Erratum: Novel optical approach to atomic force microscopy [Appl. Phys. Lett. 5 3, 1045 (1988)]. *Appl. Phys. Lett.* **1988**, *53*, 2400–2402. [CrossRef]
2. Lavrik, N.V.; Sepaniak, M.J.; Datskos, P.G. Cantilever transducers as a platform for chemical and biological sensors. *Rev. Sci. Instrum.* **2004**, *75*, 2229–2253. [CrossRef]
3. Gunter, R.; Delinger, W.G.; Manygoats, K.; Kooser, A.; Porter, T. Viral detection using an embedded piezoresistive microcantilever sensor. *Sens. Actuators A Phys.* **2003**, *107*, 219–224. [CrossRef]
4. Kooser, A.; Gunter, R.L.; Delinger, W.; Porter, T.L.; Eastman, M.P. Gas sensing using embedded piezoresistive microcantilever sensors. *Sens. Actuators B Chem.* **2004**, *99*, 474–479. [CrossRef]
5. Porter, T.; Eastman, M.; Pace, D.; Bradley, M. Sensor based on piezoresistive microcantilever technology. *Sens. Actuators A Phys.* **2001**, *88*, 47–51. [CrossRef]
6. Amírola, J.; Rodríguez, A.; Castañer, L.; Santos, J.; Gutiérrez, J.; Horrillo, M. Micromachined silicon microcantilevers for gas sensing applications with capacitive read-out. *Sens. Actuators B Chem.* **2005**, *111*, 247–253. [CrossRef]
7. Napoli, M.; Bamieh, B.; Turner, K. A capacitive microcantilever: Modelling, validation, and estimation using current measurements. *J. Dyn. Sys. Meas. Control* **2004**, *126*, 319–326. [CrossRef]
8. Bausells, J. Piezoresistive cantilevers for nanomechanical sensing. *Microelectron. Eng.* **2015**, *145*, 9–20. [CrossRef]
9. Zhao, R.; Jia, D.; Wen, Y.; Yu, X. Cantilever-based aptasensor for trace level detection of nerve agent simulant in aqueous matrices. *Sens. Actuators B Chem.* **2017**, *238*, 1231–1239. [CrossRef]
10. Patkar, R.S.; Vinchurkar, M.; Ashwin, M.; Rao, V.R. A Novel PET-Based Piezoresistive MEMS Sensor Platform for Agricultural Applications. *JMemS* **2017**, *26*, 746–748. [CrossRef]
11. Ku, Y.-F.; Huang, L.-S.; Yen, Y.-K. A real-time thermal self-elimination method for static mode operated freestanding piezoresistive microcantilever-based biosensors. *Biosensors* **2018**, *8*, 18. [CrossRef] [PubMed]
12. Xu, J.; Bertke, M.; Li, X.; Mu, H.; Zhou, H.; Yu, F.; Hamdana, G.; Schmidt, A.; Bremers, H.; Peiner, E. Fabrication of ZnO nanorods and Chitosan@ZnO nanorods on MEMS piezoresistive self-actuating silicon microcantilever for humidity sensing. *Sens. Actuators B Chem.* **2018**, *273*, 276–287. [CrossRef]
13. Xu, J.; Setiono, A.; Peiner, E. Piezoresistive Microcantilever with SAM-Modified ZnO-Nanorods@ Silicon-Nanopillars for Room-Temperature Parts-per-Billion NO2 Detection. *ACS Appl. Nano Mater.* **2020**, *3*, 6609–6620. [CrossRef]
14. Rotake, D.R.; Darji, A.; Kale, N.S. Highly selective piezoresistive sensor for mercury (Hg^{2+}) ions detection using mercaptosuccinic acid-functionalized microcantilevers with cross-linked pyridinedicarboxylic acid. *SeRv* **2020**, *40*, 543–558. [CrossRef]
15. Kandpal, M.; Behera, S.N.; Singh, J.; Palaparthy, V.; Singh, S. Residual stress compensated silicon nitride microcantilever array with integrated poly-Si piezoresistor for gas sensing applications. *Microsystem Technol.* **2020**, *26*, 1379–1385. [CrossRef]
16. Zhu, W.; Park, J.S.; Sessler, J.L.; Gaitas, A. A colorimetric receptor combined with a microcantilever sensor for explosive vapor detection. *Appl. Phys. Lett.* **2011**, *98*, 123501. [CrossRef]
17. Liu, Y.; Schweizer, L.M.; Wang, W.; Reuben, R.L.; Schweizer, M.; Shu, W. Label-free and real-time monitoring of yeast cell growth by the bending of polymer microcantilever biosensors. *Sens. Actuators B Chem.* **2013**, *178*, 621–626. [CrossRef]
18. Yu, X.; Thaysen, J.; Hansen, O.; Boisen, A. Optimization of sensitivity and noise in piezoresistive cantilevers. *J. Appl. Phys.* **2002**, *92*, 6296–6301. [CrossRef]
19. Beckwith, T.G.; Marangoni, R.D.; Lienhard, J.H. *Mechanical Measurements*; Pearson: London, UK, 2009; p. 360.
20. Temperature Coefficient of Resistance. Available online: http://www.radio-electronics.com/info/formulae/resistance/resistance-temperature-coefficient.php (accessed on 11 November 2020).
21. Allen, M. The effect of tension on the electrical resistance of single bismuth crystals. *Phys. Rev.* **1932**, *42*, 848. [CrossRef]
22. Hoffmann, K. *Applying the Wheatstone Bridge Circuit*; HBM Germany: Darmstadt, Germany, 1974; p. 25.

23. Eastman, T.; Zhu, D.-M. Adhesion forces between surface-modified AFM tips and a mica surface. *Langmuir* **1996**, *12*, 2859–2862. [CrossRef]
24. Zhao, R.; Ma, W.; Wen, Y.; Yang, J.; Yu, X. Trace level detections of abrin with high SNR piezoresistive cantilever biosensor. *Sens. Actuators B Chem.* **2015**, *212*, 112–119. [CrossRef]
25. Moulin, A.; O'shea, S.; Welland, M.E. Microcantilever-based biosensors. *Ultmi* **2000**, *82*, 23–31. [CrossRef]
26. Shekhawat, G.; Tark, S.-H.; Dravid, V.P. MOSFET-embedded microcantilevers for measuring deflection in biomolecular sensors. *Science* **2006**, *311*, 1592–1595. [CrossRef] [PubMed]
27. Zhao, R.; Sun, Y. Polymeric flexible immunosensor based on piezoresistive micro-cantilever with PEDOT/PSS conductive layer. *Sensors* **2018**, *18*, 451. [CrossRef]
28. Nan, T.; Wu, S.; Zhao, H.; Tan, W.; Li, Z.; Zhang, Q.; Wang, B. Development of a secondary antibody thio-functionalized microcantilever immunosensor and an ELISA for measuring ginsenoside Re content in the herb ginseng. *Anal. Chem.* **2012**, *84*, 4327–4333. [CrossRef]
29. Kanamori-Kataoka, M.; Kato, H.; Uzawa, H.; Ohta, S.; Takei, Y.; Furuno, M.; Seto, Y Determination of ricin by nano liquid chromatography/mass spectrometry after extraction using lactose-immobilized monolithic silica spin column. *J. Mass Spectrom.* **2011**, *46*, 821–829 [CrossRef]
30. Campos, A.R.; Gao, Z.; Blaber, M.G.; Huang, R.; Schatz, G.C.; Van Duyne, R.P.; Haynes, C.L. Surface-Enhanced Raman spectroscopy detection of ricin B chain in human blood. *J. Phys. Chem. C* **2016**, *120*, 20961–20969. [CrossRef]
31. Shyu, R.-H.; Shyu, H.-F.; Liu, H.-W.; Tang, S.-S. Colloidal gold-based immunochromatographic assay for detection of ricin. *Toxicon* **2002**, *40*, 255–258. [CrossRef]
32. Anderson, G.P.; Glaven, R.H.; Algar, W.R.; Susumu, K.; Stewart, M.H.; Medintz, I.L.; Goldman, E.R. Single domain antibody-quantum dot conjugates for ricin detection by both fluoroimmunoassay and surface plasmon resonance. *Anal. Chim. Acta* **2013**, *786*, 132–138. [CrossRef]
33. Liu, Z.-W.; Tong, Z.-Y.; Liu, B.; Hao, L.-Q.; Mu, X.-H.; Zhang, J.-P.; Gao, C. Piezoresistive microcantilever aptasensor for ricin detection and kinetic analysis. *AIP Adv.* **2015**, *5*, 041324. [CrossRef]

micromachines

Article

Experimental Validation of a Novel Generator of Gas Mixtures Based on Axial Gas Pulses Coupled to a Micromixer

Florian Noël [1,2,3], Claire Trocquet [2], Christophe A. Serra [3] and Stéphane Le Calvé [1,2,*]

1 ICPEES UMR 7515, Université de Strasbourg/CNRS, F-67000 Strasbourg, France; noel.florian.gim@gmail.com
2 In'Air Solutions, 25 Rue Becquerel, F-67087 Strasbourg, France; claire.trocquet@gmail.com
3 Institut Charles Sadron (ICS) UPR 22, Université de Strasbourg/CNRS, F-67000 Strasbourg, France; ca.serra@unistra.fr
* Correspondence: slecalve@unistra.fr

Abstract: In this work, a novel generator of gas mixtures previously numerically investigated and based on axial gas pulses coupled to a micromixer has been conceived, manufactured, and validated. Standard gaseous pollutant mixtures and pure nitrogen or pure air were introduced in a microdevice designed to generate alternating axial gas pulses which were downstream homogenized by means of a multi-stage modular micromixer. The dilution, and therefore the final pollutant concentration, was controlled by two parameters: the ratio between the times of each of the two gas pulses and the partial pressure of the pollutant(s) mixture added to the device. The gas mixture generator was coupled to an analyzer to monitor the concentration of aromatic pollutants. The response time was optimized to be lower than 2 min in accordance with the analytical instrument. The quantity of pollutants measured at the micromixer's outlet increased linearly with the expected gas concentration of 3.7–100 ppb generated by this novel microfluidic generator and fitted perfectly with those obtained by a reference gas dilution bench. At 5 ppb, the precision on the concentration generated is close to that obtained with the conventional gas mixing bench, i.e., around 10%.

Keywords: gas generator; gas mixing; pulsed flow; micromixer; multi-stages; Benzene; aromatic compounds

Citation: Noël, F.; Trocquet, C.; Serra, C.A.; Le Calvé, S. Experimental Validation of a Novel Generator of Gas Mixtures Based on Axial Gas Pulses Coupled to a Micromixer. *Micromachines* **2021**, *12*, 715. https://doi.org/10.3390/mi12060715

Academic Editor: Stefano Zampolli

Received: 8 March 2021
Accepted: 8 June 2021
Published: 18 June 2021

Publisher's Note: MDPI stays neutral with regard to jurisdictional claims in published maps and institutional affiliations.

1. Introduction

The mixing of different compounds with a known and precise ratio is often of great interest in many areas. On industrial level, this is the case for mixtures involving solids, such as in agriculture [1–3], pharmaceutical and cosmetics industries [4–7], food industry [8,9] or the manufacture of certain materials such as ceramics [10,11]. Other applications involve mixing liquids, such as for the food industry [8,12], refining fuels when choosing octane level [13,14], or managing chlorine levels in swimming pools [15]. Finally, gas mixtures are also used in many fields, for example in the beer brewing process [16], for the generation of anesthetic gases [17,18] and the direct mixing of shielding gases for coated electrode arc welding (SAEE or SMAW), but also for the generation of standard gases.

These latter cases are useful in many applications, since they can be stored in cylinders and reused for other processes [19,20] or can be directly consumed online in the case of the calibration of an analytical device [21–23]. In addition, analytical instruments with low detection limits have been in much demand for many decades. Thus, they require accurate and reliable calibrations. For example, liquid standards are used for the calibration of spectrophotometers [24,25] and High Performance Liquid Chromatography (HPLC) devices. Standard gases are mainly used for the calibration of sensors [26–28], gas chromatographs [29,30] and other fluorescence detectors [31]. This calibration, on which the reliability of the measurements made by the analytical apparatus strongly depends, is crucial and must be accurate, repeatable, and reproducible in order to guarantee the quality of measurements over time.

The Institute of Chemistry and Processes for Energy, Environment and Health (ICPEES, Strasbourg, France) and In'Air Solutions have codeveloped a Benzene, Toluene, Ethylbenzene and Xylenes (BTEX) microanalyzer [32–35]. This analyzer has several major advantages: a detection limit of 1 ppb, a low sampling flow rate of 15–50 NmL min^{-1} and an increased portability for on-site measurements (6 kg and small footprint). This last advantage is important because it allows control over the indoor air quality in public buildings, whose pollution has the greatest impact on the population health.

Unfortunately, this key point is strongly impacted by the lack of calibration devices on the market that combine precision and portability. Indeed, most analyzers are currently calibrated in the laboratory, and are not recalibrated until they return. This is mainly due to the very high dilution rates, in the range of 1 to 10 L min^{-1}, which are required to generate gas pollutant concentrations on the order of ppb, as shown in Table 1.

Table 1. Limiting characteristics of the different mixture generating devices on the market.

	Type of Compounds	Weight (kg)	Range of Gas Concentrations (ppb)	Gas Flow Rate (NmL min^{-1})	Accuracy (%)
CGS-12	Gas	21		3000	1
HovaCAL®	Liquid and gas	15	1 to 2666 *	10,000	<2
FlexStream™	Liquid and gas	12.2	<1 to 10,000	10,000	- **
491Flex™	Liquid and gas	13.6	<1 to 10,000	10,000	- **
EcoFlex™	Liquid and gas	12.2	<1 to 10,000	5000	- **
Span Pac™ I	Liquid and gas	27.2	1 to 1000	5000	- **
KinTek 491M	Liquid and gas	14.5	1 to 1000	5000	- **
Dynacalibrator®	Liquid and gas	4.8	>10 to 1000	1200	- **
Span Chek™	Liquid and gas	5.2	1 to 100	1500	- **
OVG-4™	Liquid and gas	4.5	50 to 1000	11,000	- **

* For aqueous solutions containing 0.1% w/w of compound; ** Information not given.

This leads, on the one hand, to the need for large and therefore cumbersome gas tanks, and on the other hand, to a considerable loss of compounds via a permanent leak when a lower flow rate is required for calibration.

Indeed, the most conventional method to produce pollutant concentrations ranging from 1 to 100 ppb requires the use of a gas cylinder with a concentration greater than or equal to 100 ppb, typically two commercial gas BTEX mixtures at either 100 or 1000 ppb of each compound. In this case, the method for diluting this stock gas mixture is to continuously mix a carrier gas flow rate ($Q_{carrier}$) with a pollutant flow rate (Q_{pol}) described in Figure S1a. The ratio $\frac{Q_{pol}}{Q_{carrier}+Q_{pol}}$ is then equal to the resulting dilution ratio, denoted herein as R. To achieve a concentration of 1 ppb with a carrier gas flow rate of 25 NmL min^{-1} and a mother gas mixture of 1000 ppb (i.e., R = 1000), it is necessary to be able to generate a BTEX gas flow rate of 25/1000 NmL min^{-1}, i.e., 25 NμL min^{-1}. This is just not possible with any actual gas flow regulators.

To overcome this problem, a novel gas mixture generator equipped with a micromixer, already numerically studied [36], and operating with a typical total gas flow rate of 25–50 NmL min^{-1}, has been developed in this work as part of the European LIFE Smart In'Air project. The purpose of this new gas mixture generator was to combine precision and portability, in order to build on the advantages of the aforementioned BTEX microanalyzer. Thus, it should (i) allow calibration in the range required by the analyzer, i.e., between 1 and 100 ppb; (ii) generate these concentrations at a flowrate of c.a. 25–50 NmL min^{-1}; and (iii) have a minimum size and weight (under 5 kg).

In order to achieve a continuous and very low pollutant flow rate, the gas flow can be divided into discrete pulses as illustrated in Figure S1b. Carrying out short pollutant gas trains interspersed with pure carrier gas trains allows us to achieve a dilution ratio that now depends on the ratio between the durations of the gas trains. This results in a so-called "temporal dilution", with a temporal ratio (R_{temp}) defined as follows:

$$R_{temp} = \frac{t_{pol}}{t_{pol} + t_{carrier}} \quad (1)$$

where t_{pol} is the duration of a pollutant train and $t_{carrier}$ that of a pure carrier gas train. For example, setting such t_{pol} at 1 s and $t_{carrier}$ at 9 s would allow a dilution by one decade. To help in getting a constant pollutant concentration over time, a novel micromixer was designed, fabricated and placed downstream of the gas mixture generator.

Such a configuration has been numerically studied to investigate its potential mixing capability [36]. By means of computational fluid dynamics, our research group has already demonstrated that a homogeneous gas mixture could be obtained at the outlet of the micromixer.

The present work aims at experimentally validating the above modular configuration, which makes it possible to generate gas mixtures with extremely low final concentrations (1 to 100 ppb), while using low flow rates typically varying between 25 and 100 NmL min^{-1}, unlike conventional apparatuses. Indeed, the latter usually work with a total gas flow rate of 1 to several NL min^{-1} to perform high dilution. This novel method used for the generation of gas pulses is described in detail below as well as in the calculations of the gas mixture concentrations generated and the experimental development and fabrication of a gas micromixer. The experimental validation is presented in the results section.

2. Materials and Methods

2.1. Method Used to Generate Gas Pulses

Herein, the method used to create pulses consists of a solenoid valve that alternately generates pure and polluted carrier gas trains. Such an operating procedure was described and investigated by Martin et al. [37] In their system, the gaseous pollutant mixture was generated through a permeation tube, a means of producing low flow rates of pure pollutant by evaporation, but which depends on precise control of parameters such as temperature [38]. However, the use of a commercial gas cylinder at a fixed concentration as a source of gas pollutants could alleviate the need to precisely regulated the temperature of the permeation membrane tube. In addition, the apparatus presented by Martin et al. showed a permanent outlet leak of the carrier gas when a pollutant was generated, and vice versa, resulting in a significant gas loss.

One method to eliminate this disadvantage is the one used for sample injection in gas or liquid chromatography [39,40]. This process involves a 6-way 2 position solenoid valve. The first position allows it to fully fill the sampling loop with the desired sample, while the gas or liquid mobile phase is injected into the chromatographic column and then into the detector for separation and detection, respectively. When the solenoid valve position is switched, the content of this loop is carried away by the mobile phase to the chromatography column, thus allowing the injection of a small sample volume into the mobile phase.

This analogy drove us to build a dilution device from a 6-way 2 position solenoid valve (MTV-6SL-N32UF-1, Takasago Fluidic Systems, Nagoya, Japan) as shown in Figure 1. Indeed, it is possible to create the pulses of a pollutant in a loop isolated from the continuous flow of carrier gas and then purge its content by changing the position of the solenoid valve. However, in the operation presented for liquid chromatography, the sample circulates permanently and thus requires a leak. To avoid this disadvantage, it is possible to transform the open sampling loop into a closed one. This closed loop is called hereafter "injection cell". The injection of a pollutant into this cell is then performed thanks to a 3-way solenoid valve at 2 positions, where 2 ways are part of the cell, and the third way is connected to a pressurized gas pollutant cylinder of a fixed concentration (see Figure 1).

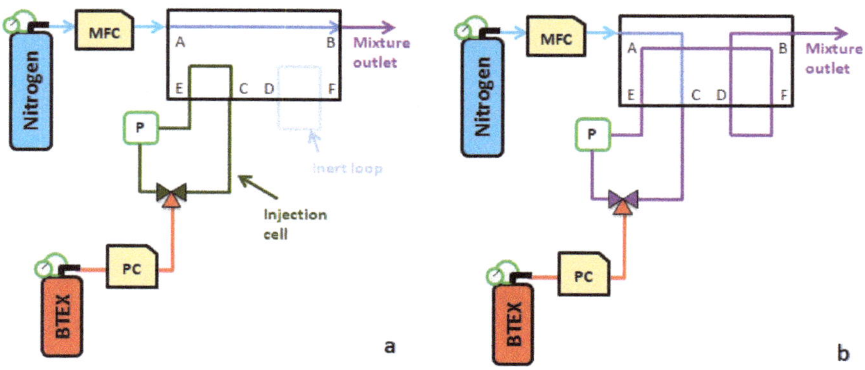

Figure 1. Schematic diagram of the gas pulse generator with a 6-way 2 position solenoid valve allowing it to generate a pure nitrogen gas train (**a**) and a nitrogen train loaded with a gas pollutant (**b**).

2.2. Calculation of the Generated Gas Mixture Concentration

The injection cell is initially filled with a pure carrier gas at near-atmospheric pressure. When the 3-way solenoid valve is in position 1, its third channel, in red in Figure 1a, is directly connected to the injection cell. Thus, due to the pressure difference between the tube located downstream to the Pressure Controller (PC in Figure 1) and the injection cell itself, the gas pollutant enters the injection cell, inducing a pressure increase. In this way, a known partial pressure of gas pollutant is added to the quasi-atmospheric pressure of pure carrier gas initially present in the injection cell. This allows to carry out a first dilution of the base mixture, called "dilution by injection", whose R_{inj} ratio is given by:

$$R_{inj} = \frac{P_{pol}}{P_{atm} + P_{pol}} = \frac{P_{pol}}{P_{cell}} \tag{2}$$

where P_{atm} and P_{pol} are the atmospheric pressures of the carrier gas already present in the injection cell and the partial pressure of the gas pollutant added into the injection cell, respectively. P_{cell} is the final pressure reached once the pollutant has been added.

It is then possible to vary the value of R_{inj} by changing the partial pressure P_{pol} thanks to the pressure controller (IQ-Flow, Bronkhorst, Montigny-lès-Cormeilles, France) (see Figure 1). Indeed, the downstream pressure imposed by the regulator gives the final value of the pressure in the injection cell. Once this pressure is reached, the 3-way solenoid valve moves to position 2, thus isolating the injection cell whose content is ready to be purged. To determine if the final pressure in the cell is the expected one, an online pressure sensor (MPS3, Elveflow, Paris, France) was used to monitor the pressure variation.

The cell is then purged in the same way as for the sample injection into the chromatography column. The 6-way solenoid valve changes position, thus redirecting the carrier gas flow towards the injection cell whose content is flushed. This forms a carrier gas train loaded with a single gas pollutant pulse. The purging time depends on 3 parameters: (i) the carrier gas flow rate $Q_{carrier}$; (ii) the total pressure of the injection cell P_{cell}; and (iii) its volume V_{cell}. Thus, for the cell to be fully flushed, the flush duration t_{flush} must be longer than:

$$t_{flush} = \frac{V_{cell}}{Q} \times \frac{P_{cell}}{P_{\text{atm}}} \tag{3}$$

In this study, the cell volume was defined based on the following parameters:

$Q_{carrier}$ = 25 NmL min^{-1}, the minimum flow rate for the mixing generator;

P_{cell} = 2000 mbar, the maximum total pressure of the cell corresponding to a two-fold dilution (1000 mbar of pure carrier gas and 1000 mbar of pollutant);

t_{flush} = 1 s, the fixed purge time of the cell.

Thus, the necessary cell volume was calculated to be equal to V_{cell} = 208.33 µL.

Once the content of the cell has been purged, the 6-way solenoid valve changes position again, allowing pure carrier gas to be generated during a new filling of the cell. Alternating pure carrier gas trains and pollutant-loaded carrier gas trains then make it possible to achieve a second dilution called "time dilution". The ratio between the pulse duration of a pure carrier gas train and the pulse duration of a carrier gas train loaded with pollutant gives the ratio of this second dilution as established in Equation (1).

By combining the dilution by injection (Equation (2)) and the temporal dilution (Equation (1)), the final dilution ratio R_{final} of the base mixture is obtained as follows:

$$R_{final} = R_{temp} \times R_{inj} = \frac{t_{pol}}{t_{pol} \times t_{carrier}} \times \frac{P_{pol}}{P_{cell}} \tag{4}$$

However, the use of a 6-way solenoid valve for this operation implies the presence of a so-called "inert" loop in the circuit, as shown in Figure 1, which contains a given volume of carrier gas at quasi-atmospheric pressure, in addition to the final dilution ratio (Equation (4)). Considering the dilution by injection, the dilution time and the content of this inert loop, we can see that the flow is composed of many gas trains whose duration and pressure are important in the calculation of the final concentration (Figure 2).

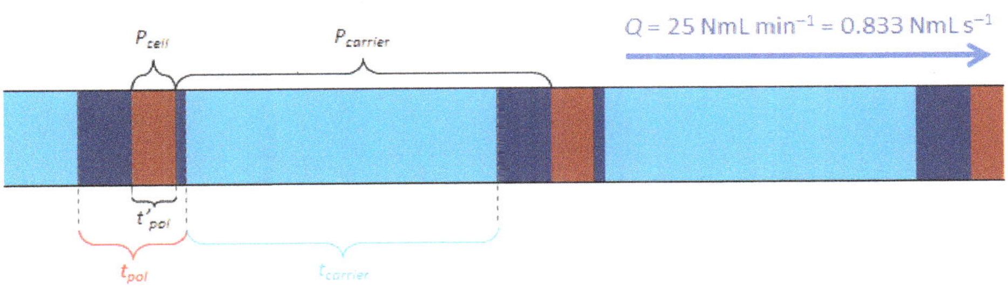

Figure 2. Schematic drawing of the flow composition and pressure exiting the mixing generator. The usual total flow rate is Q = 25 NmL min^{-1}.

Finally, the fixed duration of 1 s of a carrier gas train loaded with pollutant implies, in cases where P_{cell} is less than 2000 mbar, an excess purging of the injection cell content. It is then necessary to consider the excess of pure carrier gas generated during this duration, i.e., 1 s. This additional dilution is once again composed of pure carrier gas at near atmospheric pressure. This leads to no longer considering the duration of a polluted carrier gas train with the value t_{pol}, but with a new lower value t'_{pol} as illustrated in Figure 2. t'_{pol} is defined according to the following Equation:

$$t'_{pol} = \frac{V_{cell}}{Q} \tag{5}$$

It is thus necessary to correct t_{pol} in Equation (4) by replacing it with t'_{pol}.

In addition, it is also necessary to consider the overpressure of the cell content. Indeed, a $P_{cell}/P_{carrier}$ factor must be added to the dilution ratio to complete the equation. Finally, we obtain a flow composed of 4 distinct segments, whose characteristics are given in Figure 2 according to a schematic representation of the cycles. Taking all these data into account, the corrected final dilution ratio R'_{final} can be determined according to the following Equation:

$$R'_{final} = \frac{t'_{pol}}{t_{pol} \times t_{carrier}} \times \frac{P_{pol}}{P_{cell}} \times \frac{P_{cell}}{P_{carrier}} = \frac{t'_{pol}}{t_{pol} \times t_{carrier}} \times \frac{P_{pol}}{P_{carrier}} \quad (6)$$

2.3. Chemicals

The BTEX concentrations were generated from two different standard BTEX mixtures purchased from Messer (Folschviller, France) with an initial concentration of overall compounds equal to either 100 ppb or 1000 ppb and with a 10% uncertainty for both gas cylinders. Both standard mixtures were diluted with pure air or pure nitrogen (99.999% purity, Messer, Folschviller, France) to obtain the targeted concentrations of BTEX mixtures.

2.4. Measuring the Generated Gas Mixture Concentrations

The apparatus used for the analysis of the concentration produced by the generator of mixtures is a BTEX analyzer (Benzene, Toluene, Ethylbenzene and Xylenes) provided by the company In'Air Solutions (µBTEX-1, Strasbourg, France). This device is a portable and compact Gas Chromatograph (GC), which allows these compounds to be separated to measure their gas phase concentrations.

The working principle of this analyzer, developed by Nasreddine et al. in 2015 [34], has been widely described in recent publications [33,35]. Analyzing a sample consists of 3 steps: sampling, separation, and detection. During sampling (Figure S2a), a pump (SP570.ECBLa, Schwarzer Precision GmbH, Essen, Germany) makes the sample to flow through a 6-way/2 positions valve and a sampling loop of 200 µL. The flow rate of the sample is limited by an external Mass Flow Controller (MFC) (IQ-Flow, Bronkhorst, France) placed upstream to the pump.

The sample passes continuously through the 200 µL-loop. In parallel to this sampling, a flow of pure nitrogen, used as carrier gas, flows through a capillary column and then through a Photo-Ionization Detector (PID) (piD-TECH eVx blue, Baseline, Baseline-MOCON Inc., Lyons, GA, USA) at a flow rate of 2.5 NmL min^{-1}, this flow rate being regulated by a Pressure Controller (PC) (IQ-Flow, Bronkhorst, Montigny-lès-Cormeilles, France) placed upstream to the 6-way valve.

The second step, consisting of separating the compounds of the sample, starts by switching the 6-way valve (Figure S2b). The gaseous sample inside the 200 µL-loop is then flushed by the carrier gas and injected into the capillary column. Here, the compounds are separated depending on their affinity with the solid phase of the chromatographic column (internal diameter (ID) 0.18 mm, RXi-624 stationary phase, 1 µm film thickness, Restek, Bellefonte, PA, USA), which differs according to each compound's physical properties. Indeed, Benzene is less retained by the column while o-Xylene is longer adsorbed on the capillary column walls.

After exiting the column, each compound passes through the photoionisation detector (PID) where it is quantified for the last step by means of an external calibration performed prior the experiments. For BTEX, compounds exit the capillary column in this order from first to last: Benzene, Toluene, Ethylbenzene, m- and p-Xylenes (co-eluted) and o-Xylene. The signal provided by the PID is then obtained in µV: the more molecules detected, the higher is the signal. The concentration of a compound can then be determined by the area under its related signal peak.

2.5. Experimental Development and Fabrication of a Gas Micromixer

In order to ensure a fast response time and the highest possible homogeneity to the gas flow leaving the above gas mixture generator, a microfluidic homogenization device had to be developed [36]. So far, axial gas heterogeneities have been poorly represented and investigated. Martin et al. (2012) proposed a homogenization method, but it was limited because it was specifically designed for their application and a given time dilution ratio, which did not match with our need for a flexible homogenization method. However, in the case of the gas mixture generator presented here, achieving the desired concentration range requires the ability to modify this dilution ratio.

Thus, a microfluidic homogenization circuit was investigated. The experimental development reported hereafter was supported by a previous theoretical study conducted by Computational Fluid Dynamics (CFD) [36]. In the case of gas mixtures, it should be noted that the diffusion coefficient D between gases (0.1 $cm^2\ s^{-1}$) is significantly higher than that between liquids (1.10^{-5} $cm^2\ s^{-1}$), i.e., typically four orders of magnitude higher. In fact, it becomes very interesting to take advantage of this natural phenomenon for mixing gas trains. To achieve this, it is necessary to reduce the value of the Peclet number (Pe) for the gas stream studied. Indeed, this number represents the ratio of a mass transport rate by convection and by diffusion:

$$Pe = \frac{L_c \times v}{D} \quad (7)$$

where L_c is a characteristic length (cm), v is the local gas velocity (cm s^{-1}) and D is the mass diffusion coefficient ($cm^2\ s^{-1}$).

In order to give more importance to mass transport by diffusion, it becomes necessary to reduce the convection term and thus to reduce the average velocity v of the fluid. The concept of a buffer tank allows us to achieve this objective by increasing the cross-section through which the gas mixture flows. Thus, the larger the cross-section of the buffer tank compared to the upstream channel, the more homogenous the mixture will be.

To ensure an efficient mixing at a flow rate of 5 NmL min^{-1}, a buffer tank was considered to mix a pure 9 s carrier gas train with a 1 s pollutant-loaded carrier gas train, i.e., a volume of 0.833 mL. These conditions correspond to the extreme case of the "temporal dilution" in the present work. Consequently, it was arbitrarily chosen to build a tank to hold 1.5 times this volume, i.e., about 1.2 mL. This first tank had the dimensions 30 × 40 × 1 mm (length × width × depth).

However, a section too large may involve dead zones whose volumes are not renewed (i.e., flushed), thus impacting the response time of the overall circuit. To avoid this phenomenon as much as possible, it was chosen to rely on 4 smaller buffer tanks placed in parallel. It thus offers the same overall cross-section increase while limiting possible dead zones and offering an additional benefit arising from the 4-fold reduction of the convective mass transport. We then obtain a pattern of 4 tanks, each with dimensions of 30 × 10 × 1 mm (Figure 3a and Figure S3A).

Finally, in order to meet the need for homogenizing mixtures generated at flow rates equal or above 25 NmL min^{-1}, we considered modular gas micromixers composed of the assembly of X-alternated rectangular buffer stages (Figure 3a) and connection stages, X being the number of buffer stages, i.e., the number of mixing plates as defined in Figure 3. Figure 3b represents the assembly of a 4-stage micromixer needed to homogenize the axial heterogenous gas mixture at a flow rate of 25 NmL min^{-1}. Each buffer stage (79 × 112 × 10 mm) comprises 1 pattern of 4 buffer tanks with their inlet and outlet network of microchannels (1300 × 1300 µm) (Figure 3b). Connection stages of a lower thickness (5 mm) comprise transversal microfluidic channels (1300 × 1300 µm) for connecting two buffer stages or to fix a 1/16" fluid connector. Thus, the entire assembly represents a multi-stage micromixer.

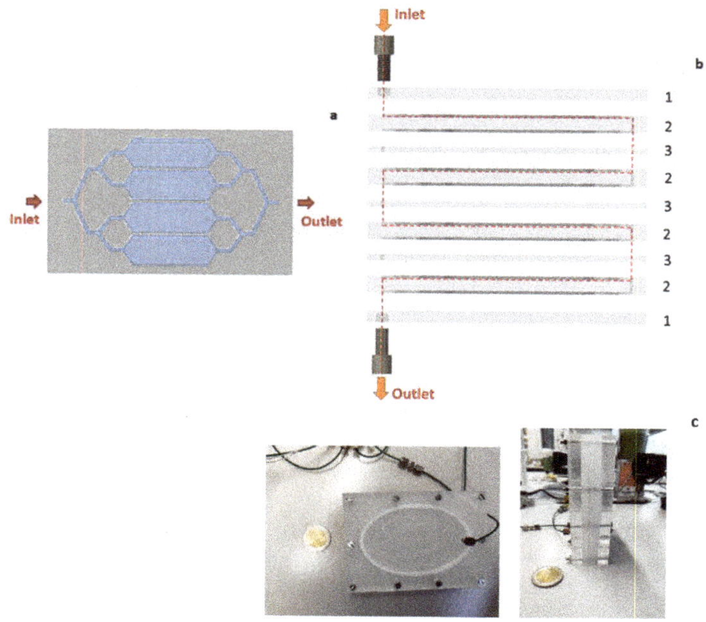

Figure 3. Schematic drawing of (**a**) the design of a 4 buffer stage as numerically studied by Noel et al. (2019) [36] for which dimensions can be found in Figure S3A; (**b**) image of a 4-stage micromixer [36] comprising four mixing plates milled on one side with a four buffer tanks (2), one inlet and outlet plate (1) as well as one connection plate (3) to link two mixing plates; (**c**) photographs of the 4-stage micromixer prototype manufactured and tested in this work.

A laboratory prototype of this micromixer was developed and tested as part of this study (Figure 3c). It was made of Poly(methyl methacrylate) (PMMA), a material inert to the pollutants studied, and the patterns were engraved using a micro-milling machine (Roland MDX-40A, Serris, France). The connectors used for the mixer inlet and outlet tubes were 1/16″ Polyetheretherketone (PEEK) connectors. Sealing between the stages was ensured by 3 mm diameter Teflon O-rings, inserted into grooves, and compressed with screws passing through the entire mixer.

In order to manufacture the stages of the micromixer faster, two different milling tools were used (Figure S3B). At first, the raw shape of the 4 buffer zones pattern, the screw paths and O-ring housing, were conducted with a 1.5 mm milling tool (green on Figure S3B). Once this step was achieved, the microchannels were manufactured using a 0.5 mm milling tool (red on Figure S3B), leading to satisfactory precision.

3. Results

The objective of this work is to compare the experimental results obtained with the gas mixture generator coupled or not to the gas micromixer described above and check the consistency with the results determined numerically by Computational Fluid Dynamics (CFD) in a previous work [36]. The purpose of adding a micromixer was to ensure a rapid homogenization of the gas train generated as illustrated in Figure 4. CFD simulations carried out in a previous study [36] highlighted a homogenization capacity of more than 95% for flow rates between 25 and 100 NmL min^{-1}. These performances were achieved with a 4-stage micromixer operating at a flow rate of 25 NmL min^{-1}, 8 stages for 50 NmL min^{-1} and 16 stages for 100 NmL min^{-1}. These numerical results indicated a linear trend between the number of stages needed, and the gas flow rate applied.

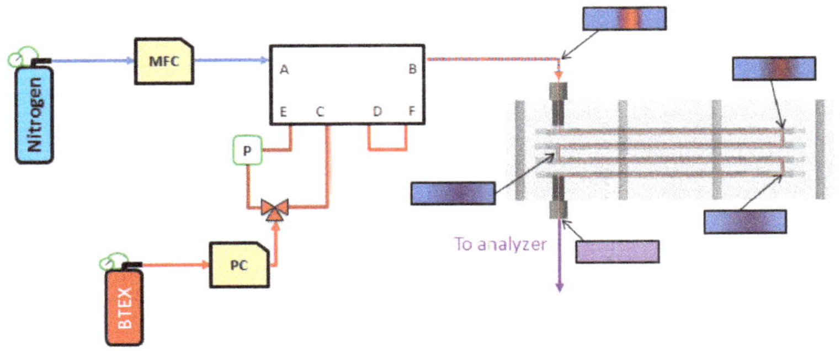

Figure 4. Illustration of the setup after adding a 4-stage micromixer to the original gas mixture generator. The inlet and outlet plates (plates 1 in Figure 3b) as well as the connection plates (plates 3 in Figure 3b) to link the mixing plates are not displayed for clarity. Insets present the evolution of the pollutant heterogeneity through the micromixer as determined by a previous simulation work [39], red being the polluted sample, blue the pure carrier gas and purple the perfect mixture of the two.

Consequently, the present experimental work was limited to the realization and study of a mixture gas generator equipped with a 4-stage micromixer operating at 25 NmL min^{-1}.

3.1. Gas Mixture Generator Alone or with a Dowstream Micromixer

First, experiments were performed using the gas mixture generator alone. A representative gas mixture at 50 ppb of each BTEX was thus chosen and generated from a commercial gas cylinder of 1000 ppb of each compound, at a flow rate of 25 NmL min^{-1}. This targeted concentration of 50 ppb was reached with P_{pol} = 200 mbar and $t_{carrier}$ = 1 s, i.e., an estimated final concentration of $C_{final} = C_{1000ppb} \times R'_{final} = 49.92$ ppb, where $C_{1000ppb}$ is the concentration of the commercial BTEX gas cylinder.

As shown in Figure 5, the response time of the generation device was quite long, in the order of several tens of minutes. Indeed, the concentration of each compound stabilized after approximately 60 min and was, however, very close to the target value of 50 ppb, which constituted a first proof of concept. However, the concentrations were all slightly below 50 ppb, which was explained by the fact that the cell pressure P_{cell} did not reach the desired value: a smaller amount of pollutant was then introduced during the dilution by injection.

One also noticed that the final benzene concentration of around 70 ppb was significantly higher than the expected 50 ppb. This was explained by a defect in the calibration of this compound, which was subsequently corrected. Despite this fact, benzene had a response time close to that of other compounds, and an equivalent stability once the steady-state regime had been established.

To solve this issue related to the response time, the optimization of the tube lengths was carried out in the whole experimental device before carrying out more tests. The aim of this optimization was to drastically reduce the total volume of the circuit and to limit dead volumes as much as possible. The total volume was thus divided by approximately 10, from more than 80 mL down to 7.4 mL.

In addition, the analysis method was modified to obtain closer measurement points over time. Thus, the chromatography column was removed to let the mixture fed directly the PID detector, thus offering a measurement point every 2 min instead of every 10 min. In fact, the separation of the compounds required approximately ten minutes, which determined the time resolution, i.e., the time between two analyses as soon as the chromatographic column was present.

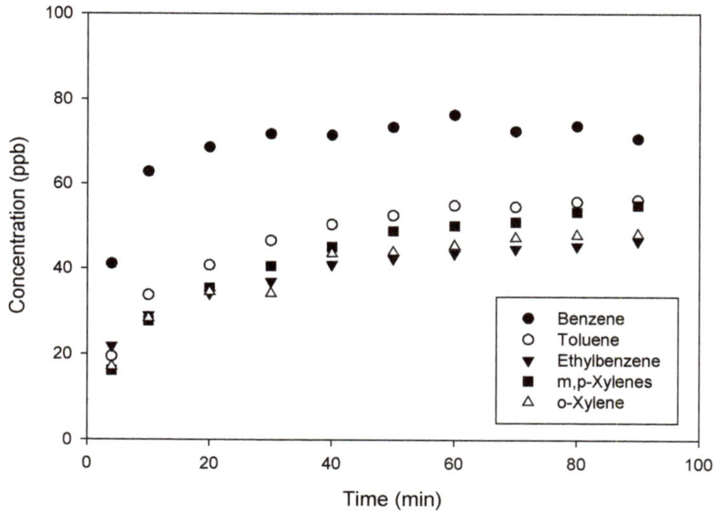

Figure 5. BTEX concentration time profiles were obtained without any micromixer by generating a targeted gas mixture of 50 ppb from a gas cylinder of BTEX at 1000 ppb of each compound. Other parameters were fixed at: Q = 25 NmL min^{-1}; P_{pol} = 200 mbar; $t_{carrier}$ = 1 s. For m,p-Xylenes, the concentration reported corresponds to the individual concentration of each compound.

In the absence of the column, the compounds were no longer separated, which made it possible to determine an overall concentration of the BTEX injected by considering that the response factor of these 6 species was similar, as reported in the literature [41]: 0.780 (Benzene); 0.794 (Toluene); 0.818 (Ethylbenzene); 0.812 (m, p-Xylenes); 0.840 (o-Xylene). This assumption was verified experimentally later since the total concentration generated was close to the targeted one.

New tests were then performed with the new configuration with a lower dead volume, generating BTEX mixtures at 20, 50, 10, and 100 ppb, consecutively (Figure 6a). A significant improvement in the response time was then observed, since the latter was reduced down to 4 min. On the other hand, the stability of the concentration was not ensured, the concentration of 10 ppb presenting a significant variation in concentration over time. The maximum and minimum areas obtained for this concentration of 10 ppb from each of the BTEX were equal to 2.29 × 10^5 a.u. and 2.05 × 10^5 a.u., respectively, which represented a variation of about 10%.

Therefore, even if the response time was considerably improved thanks to the reduction in the volume of the circuit, the instability of the concentration nevertheless required the use of a micromixer placed after the gas train generation in order to reduce the BTEX concentration fluctuation over time. This periodic instability (see the lowest concentration of 10 ppb in Figure 6a) also confirmed the axial heterogeneity already observed in our previous simulation work (Noel et al., 2019) [36]. The fact that these periodic variations did not appear for the concentrations other than 10 ppb came from the highest temporal dilution of 1/10 used in this case (t_{pol} = 1 s, $t_{carrier}$ = 9 s), while other concentrations were produced with the smaller temporal dilution of 1/2 or 1/5, implying less axial heterogeneity.

In order to confirm the numerical results, the same tests were carried out experimentally at 25 NmL min^{-1} by generating BTEX mixtures of 20, 50, 10, and 100 ppb from a BTEX gas cylinder at 1000 ppb, and by measuring the total concentration of BTEX generated at the outlet of the micromixer. The results are presented in Figure 6b and show a further significant improvement in the response time, which now appears to be less than the 2 min required for the measurement. In addition, the variations in concentration previously observed for 10 ppb due to poor homogenization in the absence of the micromixer have completely disappeared, with no periodic variations in the concentration at the micromixer outlet being visible.

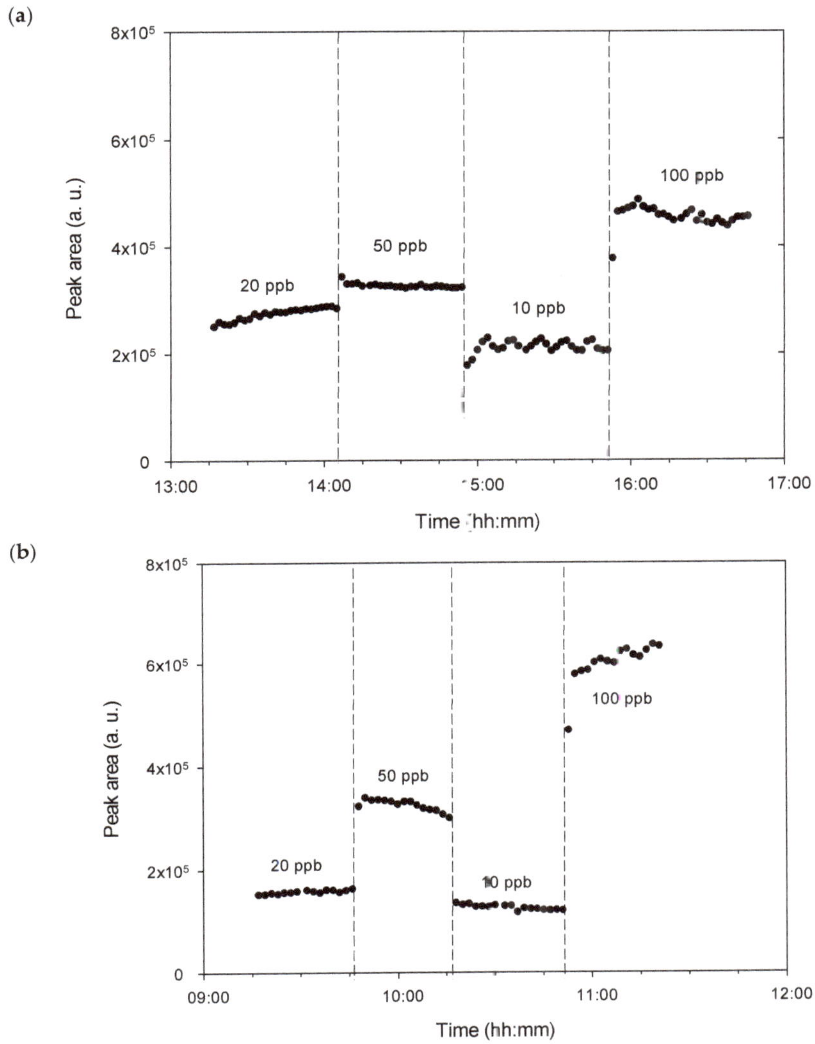

Figure 6. Total BTEX peak area on a logarithmic scale vs. time obtained either (**a**) without any micromixer or (**b**) with the gas train generator coupled to a 4-stage micromixer for gas mixtures of 20, 50, 10 and 100 ppb generated from a gas cylinder of BTEX at 1000 ppb of each compound. Concentrations of 20, 50, 10 and 100 ppb were generated with respectively P_{pol} = 200 mbar and $t_{carrier}$ = 4 s, P_{pol} = 200 mbar and $t_{carrier}$ = 1 s, P_{pol} = 200 mbar and $t_{carrier}$ = 9 s, and P_{pol} = 400 mbar and $t_{carrier}$ = 1 s. The total flow rate was fixed at Q = 25 NmL min^{-1}.

Furthermore, in the presence of the micromixer (Figure 6b), it was not possible to observe the oscillations predicted by the simulation (Figure 6 shown in our numerical paper [36]) because the time step of the measuring instrument was 2 min, whereas the period of the oscillations observed numerically was only about 10 s.

3.2. Calibration and Reproducibility Experiments

The reproducibility of the novel gas mixture generator was then verified by overlapping several series of measurements carried out under different conditions and spaced over time (see Figure 7). Similar measurements performed with a laboratory validated calibration bench were also added to compare the data obtained with the new microdevice with a reference generation source so-called dilution bench.

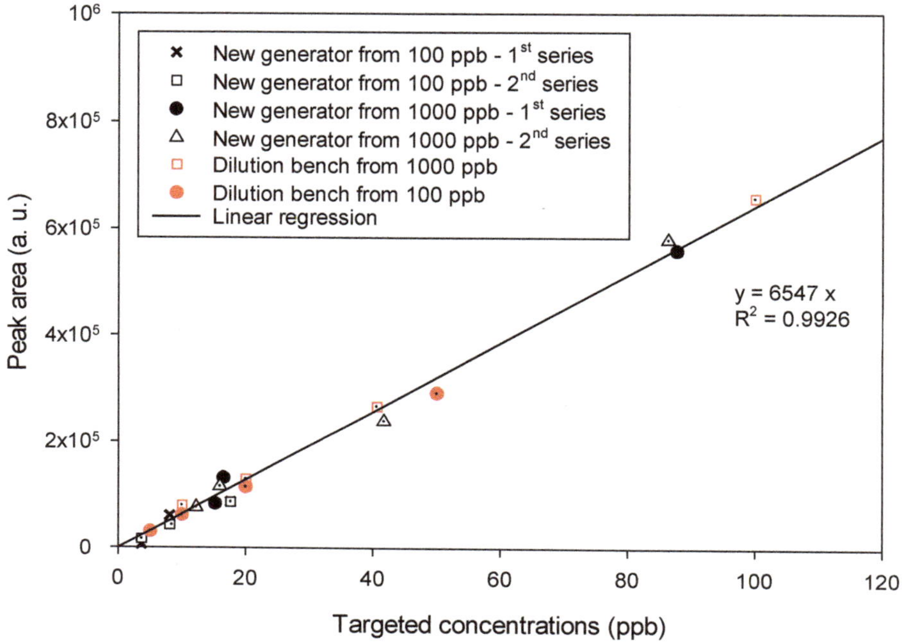

Figure 7. Comparison of the results obtained with the dilution bench and the novel source of gas mixtures based on a gas train generator coupled to a 4-stage micromixer, both using a gas cylinder of BTEX at 100 or 1000 ppb of each compound. The plot represents the experimental chromatographic peak area vs. the targeted concentration. The red symbols correspond to the data obtained with the dilution bench while the black ones were obtained with the experimental generator equipped with a 4-stage micromixer. The flow rate was fixed at: $Q = 25$ NmL min^{-1}. P_{pol} varied from 100 to 2000 mbar and $t_{carrier}$ varied from 1 to 9 s.

For each series of measurements, areas of the peaks obtained under conditions for the generation of variable BTEX mixtures were corrected by subtracting the average value of the peak obtained with pure nitrogen for 1 h after stabilization. This theoretically allows us to obtain a straight line passing through the origin when plotting the variations of areas as a function of pollutant concentration. These concentrations were recalculated from the exact experimental pressure data recorded using the pressure sensor.

The concentrations targeted and obtained are listed in Table S1 for the different series of experiments performed at $Q = 25$ NmL min^{-1}, combining both generation and measurements.

3.3. Experiments at Different Total Flow Rates

A last experiment was carried out to compare the gas mixture generation of the same concentration at different flow rates. Indeed, the generator of gas mixtures needs also to be able to operate at different flow rates to allow the calibration of a large range of analytical instruments such gas analyzers and gas sensors. Thus, flow rates of 12.5, 25 and

50 NmL min^{-1} were investigated to obtain a fixed target BTEX concentration of 50 ppb (Figure S4).

4. Discussion

4.1. Gas Mixture Generator with Dowstream Micromixer

A thorough analysis of Figure 6b shows that, for every given concentration generated, the obtained signal slightly increased or decreased depending on the concentration considered. This is because the total pressure in the cell varies around the target pressure P_{cell}, indeed the pressure regulator used did not close completely when the pressure of its setpoint was reached. Therefore, and since the pressure upstream to the regulator was always higher than the set pressure, it continued to allow the BTEX mixture to pass through it, the pressure downstream then varying regularly.

An accurate control of the pressure would make it possible to a achieve better precision and stability of the concentration generated. To fix this problem, an assembly coupling a proportional solenoid valve to a pressure sensor could be considered in the future, since this type of solenoid valve can be completely closed to stop the arrival of a pollutant train when the desired pressure P_{cell} is reached.

However, this concentration variation, due to the pressure management, was nevertheless corrected post-generation by calculating the true concentration generated. Indeed, the software developed for the control of the gas mixture generator allowed us to record and reprocess the data from the pressure sensor. It was thus possible to know at each generation cycle the real pressure introduced into the cell (P_{cell}) and therefore to calculate the exact concentration generated using Equation (6).

Even if the concentration varied compared to that of the set value entered in the software controlling the gas mixture generator, it was possible to know the final concentration generated. The accuracy of the pressure sensor being 0.2% of full scale and its maximum pressure being 7 relative bars, it was possible to detect pressure changes in the order of 14 mbar. This was equivalent to a potential error of +/− 7% for a pressure P_{pol} = 200 mbar used to obtain a gas mixture containing 50 ppb of each compound.

This error could be reduced by selecting another sensor from the same series available from the same manufacturer, such as an MPS3 pressure sensor, whose maximum relative pressure is 2 bars. Indeed, the mixture generator should not exceed a cell pressure P_{cell} of 2000 mbar absolute. The accuracy of this sensor also being 0.2% of full scale, the potential error would be reduced down to +/− 2%, always for a pressure P_{pol} = 200 mbar.

Concerning this concentration variation phenomenon, it also appeared in another experiment, during which a zero concentration (pure nitrogen) and a concentration of 50 ppb of each BTEX were alternately generated. Figure S5 shows that the repeatability of the concentration generated is very satisfactory. Indeed, Figure S5 exhibits that the two generation periods of 50 ppb are extremely similar to each other, both in terms of the final value reached and in the shape of the curve.

For the four targeted concentrations of 20, 50, 10 and 100 ppb of BTEX (Figure 6b), the mean peak areas were respectively 15.7 × 10^4 a.u., 32.5 × 10^4 a.u., 12.6 × 10^4 a.u. and 60.9 × 10^4 a.u. These mean values were respectively linked to standard deviations of 0.3 × 10^4 a.u., 1.1 × 10^4 a.u., 0.5 × 10^4 a.u. and 1.8 × 10^4 a.u. The calculated standard deviations represent thus between 2.1% and 4.1% of deviation around the mean value, depending on the generated concentration.

According to Figure 6b, a slight increase of the PID signal was observed for the concentration of 50 ppb after reaching the equilibrium regime. After stabilization, the variations between the lowest and highest value of a 50 ppb BTEX generation series for 1 h were in the order of 10%. However, due to the pressure variation at the injection cell level, this error could be corrected by recalculating it using the recorded data from the pressure sensor to know the exact concentration generated during each cycle.

In Figure S5, it could be also observed that several tens of minutes were necessary for the PID signal to stabilize when pure nitrogen was injected alone. This continuous flow of pure nitrogen was injected by the same path as the pure nitrogen pulses produced when generating a polluted mixture. For this purpose, the 6-way valve did not switch during pure nitrogen generation. It also ensured that the flow of pure nitrogen was isolated from the BTEX injection valve. This was an additional security measure to prevent unwanted BTEX amounts from getting through the valve and polluting the flow of pure nitrogen.

Since the response time of the gas mixture generator coupled with the gas micromixer was less than 2 min when switching from one concentration to another, as seen previously in Figure 6b, the memory effect due to potential adsorption on the PEEK device and Teflon tubes did not seem to be an issue. However, the renewal time of the PID detector cell could support, at least partially, these observations; the BTEX molecules being able to adsorb onto the surface of the porous membrane located at the detector inlet.

Being exposed to pure nitrogen, the peak area of the PID detector stabilizes at around 6×10^4 a.u in Figure S5. This blank value defining the baseline, when the BTEX analyzer was configured as a chromatograph equipped with a GC column, may vary slightly between two series of measurements spaced over time (several days or weeks for example).

4.2. Calibration and Reproducibility Experiments

Figure 7 groups together all the results obtained from the different series of measurements made with two different BTEX cylinders at 100 or 1000 ppb respectively combined with the dilution bench or the new generation source equipped with the micromixer. The peak area of total BTEX increased perfectly linearly with the concentration of BTEX in the range 3.7–100 ppb, the linear regression of the dataset allowed to obtain a straight line with a coefficient of determination $R^2 = 0.993$. In addition, the data obtained with the new generation source agreed perfectly with those obtained by the reference method, namely the dilution bench.

This confirms the ability of the new microfluidic gas mixture generator to produce a reliable linear calibration plot in a concentration range of 5 to 100 ppb, with initial standard BTEX gas mixtures of different concentrations. In addition, the fact that the measurements were taken between several days and 2 weeks apart also made it possible to validate the reproducibility of the gas mixture generation thanks to the new microdevice.

The standard deviation on the peak area was equal to about 1.70×10^4 a.u., which indicates an uncertainty greater than 10% for concentrations less than or equal to 20 ppb. Indeed, the area obtained from the PID signal is less than 130,000 units for this concentration. Therefore, a calibration curve performed at low concentrations should contain many calibration points to ensure a reliable linear plot.

It is not certain, however, that this uncertainty was only due to the gas mixture generator. It is possible that the PID signal also included an error in the measurement, but also that the reprocessing of the data and particularly the integration of the peak, implied another error adding to measurement and generation of uncertainties. In this case, however, it was not possible to determine the impact of each contribution on the final error.

4.3. Experiments at Different Total Flow Rates

The results presented in Figure S4 show that the PID signal was stable for each flow rate investigated, indicating that this parameter did not influence the homogeneity and the stability of the concentration generated. On the other hand, a difference in PID signal was observed for the different flow rates. For a flow rate of 50 NmL min^{-1}, the average value obtained was around 207,000 a.u., compared to 195,000 a.u. for 25 NmL min^{-1} and 148,000 a.u. for 12.5 NmL min^{-1}.

The significant difference between the above value for a flow rate of 12.5 NmL min^{-1} and those obtained with the highest flow rates is most likely due to this low total gas flow rate, which no longer allowed the complete purging of the injection cell. Indeed, the volume of this latter has been sized to allow a total purge at 25 NmL min^{-1}, which means that at a flow rate of 12.5 NmL min^{-1}, only half of the 208 µL cell was purged. As a result, only half of the BTEX contained in the injection cell was flushed during the purge step, theoretically dividing the final concentration by 2.

The small difference between the signal obtained at 25 and 50 NmL min^{-1} in Figure S4 is probably due again to the poor management of the cell pressure (P_{cell}). Indeed, the generation of an identical concentration at a flow rate of 25 NmL min^{-1} and at a flow rate of 50 NmL min^{-1} was obviously done by doubling the P_{cell} value. However, a different cell pressure implies that the pressure regulator should be capable of managing the exact value of the pressure downstream.

The pressure regulation problems are illustrated in Figure S6. Here, two series of data are recorded using the cell pressure sensor during the generation of 50 ppb of BTEX with P_{pol} = 200 mbar, $t_{carrier}$ = 1 s and a flow rate of 25 NmL min^{-1}. The value of the relative pressure P_{atm} must be a quasi-atmospheric pressure. This is relatively constant, and its average value over the entire generation time can be compared to the relative pressure P_{cell} value to determine the pressure P_{pol} of each cycle, allowing the calculation of the exact gaseous mixture concentration generated.

According to Figure S6, the relative pressure P_{atm} was quite stable, with an average of 97.5 mbar and a standard deviation of 7.34 mbar, which corresponds to a variation of 7.5%. On the other hand, this figure highlights the problem of regulating the pressure P_{cell} during the cycles. After a stabilization period where the cell pressure decreased from 550 mbar (setpoint from the previous generation experiment) to approximately 280 mbar (new setpoint), an oscillation of P_{cell} was observed between 269 and 293 mbar with a period of about 34 s, leading to an average value of 281 ± 12 mbar, i.e., a variation of 4.3%.

This implies that at a flow rate of 25 NmL min^{-1}, about 14.17 mL of the gas mixture flowed in a period of pressure oscillation. However, the BTEX analyzer (μ-BTEX1, In'Air Solutions) coupled to the generation source in this work had a sampling loop of only 200 μL. The cell pressure variation can thus strongly impact the detected concentration. Nevertheless, some of these variations were potentially damped by the micromixer during homogenization. Indeed, the volume of this device has been dimensioned to homogenize 10 s of mixture ($t_{carrier}$ = 9 s), with, in addition, an oversizing by a factor of 1.5. Therefore, in the future, this phenomenon should be investigated in detail in order to evaluate the impact on the final concentration, which could be measured by online mass spectrometry.

5. Conclusions

A new gas mixture generator was successfully conceived, designed, and manufactured to reach a dilution ratio up to 100 at a low flow rate of 25 NmL min^{-1}. This total gas flow rate is about 40 to 200 times less than the operating flow rate of commercial calibration devices. Thus, this microfluidic generator allows generating gas concentrations as low as 10 ppb of a BTEX mixture from a 1000 ppb commercial cylinder. Its low carrier gas consumption eliminates the need for large, pressurized cylinders that limits the portability of the current equipment. For example, a 58 L-gas cylinder of nitrogen weighs only 1.2 kg and allows a continuous gas generation for 38.7 h when a full gas calibration of a gas analyzer with five known concentrations needs 5 times 10 min, i.e., a total time of 50 min. In addition, the on-site calibration of analytical instruments requires only 1 or 2 calibration points if no drift with the previous full calibration is observed. In addition, the low carrier gas consumption avoids the unnecessary loss of costly gas. The weight of the developed generator itself should not exceed 5 kg and was composed only of few solenoid valves, gas connectors and a gas micromixer placed downstream to the gas train generator.

The gas homogenization was achieved using a multi-stage microfluidic mixer and was higher than 95% for a response time of less than 2 min. The design of this new micromixer, which has been also validated by previous numerical study [36], allows homogenizing gas mixtures with axial heterogeneity, which was scarcely studied until now in the literature but seems to be of great potential in view of the results obtained during the present study. In addition, this micromixer has a modular design, which allows adding new mixing plates for increasing homogenization efficiency. These plates were manufactured by a simple etching process with a lab micro-milling machine but could be also manufactured by other microfluidic microfabrication techniques including 3D printing.

Finally, the performances of this gas mixture generator could be further improved by ensuring a better pressure regulation when BTEX mixture is injected into the injection cell of the generator. Indeed, the major source of errors on the concentration generated comes from a potential regulatory failure, which could easily be solved with another technology. For example, one could couple a pressure sensor and a proportional solenoid valve to replace the pressure regulator. Such an improvement could also further reduce the production cost and the low concentration limit of the gas mixture generator: it would then be possible to inject an even lower partial pressure of BTEX into the cell, which will further increase the dilution ratio per injection. As such, it would be possible to achieve a dilution ratio of up to 1000, and to reach concentrations of around 1 ppb of BTEX from a commercial cylinder of 1000 ppb.

In the future, to address the high cost of O-rings used in the assembly of the current micromixer, consideration will be given to fabricate the micromixers in one piece by 3D printing with BTEX non-emissive and non-adsorbing polymer materials. In addition, this type of gas generator could be used for gases other than BTEX, such as for other volatile organic compounds (alkanes, aldehydes, ketones, etc.) or even gaseous inorganic pollutants (oxides of nitrogen, sulfur compounds, carbon oxides, etc.) For example, this new tool could be used for calibrating NO and NO_2 analyzers, or CO and NH_3 sensors, etc.

6. Patents

The following patent resulting from the work reported in this manuscript has been submitted:

F. Noel, S. Le Calvé, C. Serra, Microfluidic generator for generating a gas mixture. WIPO Patent Application WO/2020/178022, 10 September 2020.

Supplementary Materials: The following are available online at https://www.mdpi.com/article/10.3390/mi12060715/s1, Figure S1: Comparison between (a) the common gas dilution method and (b) the pulse dilution method., Figure S2: working principle of the μBTEX-1 (In'Air Solutions, Strasbourg, France) with (a) the sampling step and (b) the separation and detection steps, Figure S3: (A) Main dimensions (in mm) of the gas micromixer (Reproduced with permission from Noël, F.; Serra, C.; Le Calvé, Micromachines; published by MDPI, 2019), channels' depth was 1 mm, (B) schematics of the parts manufactured using a 1.5 milling tool (green) and a 0.5 mm milling tool (red), Table S1: List of targeted concentrations and those generated ranging between 3.7 and 100 ppb for 6 series of measurements performed from either a dilution bench validated in the laboratory or the new microfluidic generator of gas mixtures, Figure S4: Total BTEX peak area on a logarithmic scale vs. time obtained with the gas train generator coupled to a 4-stage micromixer for a generated gas mixture of 50 ppb at different flow rates, i.e., 12.5, 25 and 50 NmL min^{-1}. A standard gas cylinder of BTEX at 1000 ppb of each compound was used. Other parameters were fixed at: Q = 25 NmL min^{-1}; P_{pol} = 200 mbar; $t_{carrier}$ = 1s, Figure S5: Total BTEX peak area vs. time obtained with the gas train generator coupled to a 4-stage micromixer for either a gas mixture of 50 ppb generated from a gas cylinder of BTEX at 1000 ppb of each compound, or pure nitrogen, both injected through the same fluidic circuit. Other parameters were fixed at: Q = 25 NmL min^{-1}; P_{pol} = 200 mbar; $t_{carrier}$ = 1 s, Figure S6: Relative pressure-time profiles in the cell after filling with the BTEX mixture (P_{cell}) and after purge with pure nitrogen (P_{atm}). Generation parameters were fixed at: Q = 25 NmL min^{-1}; P_{pol} = 200 mbar; $t_{carrier}$ = 1 s and C_{pol} = 50 ppb.

Author Contributions: Conceptualization, F.N., S.L.C., C.A.S.; methodology, F.N., S.L.C.; software, F.N.; validation, F.N., C.T., S.L.C., C.A.S.; formal analysis, F.N., C.T.; investigation, F.N.; resources, S.L.C., C.A.S.; writing—original draft preparation, F.N., S.L.C., C.A.S.; writing—review and editing, F.N., S.L.C., C.A.S.; visualization, F.N.; supervision, S.L.C., C.A.S.; project administration, S.L.C.; funding acquisition, S.L.C. All authors have read and agreed to the published version of the manuscript.

Funding: This research was funded by "Association Nationale de la Recherche et de la Technologie," grant number 2016/1089, by European Commission through the LIFE program, grant number LIFE17 ENV/FR/000330 and by SME In'Air Solutions.

Institutional Review Board Statement: Not applicable.

Informed Consent Statement: Not applicable.

Data Availability Statement: The data presented in this study are available on request from the corresponding author. The data are not publicly available because the authors want to keep priority for conference presentations.

Acknowledgments: This study was supported by the CIFRE program between SME In'Air Solutions (Strasbourg, France) and the French National Association of Research and Technology (ANRT, grant number 2016/1089), and was also funded by the European Commission through the LIFE program (SMART'IN AIR, LIFE17 ENV/FR/000330).

Conflicts of Interest: The authors declare no conflict of interest.

References

1. Lindley, J.A. Mixing Processes for Agricultural and Food Materials: 1. Fundamentals of Mixing. *J. Agric. Eng. Res.* **1991**, *48*, 153–170. [CrossRef]
2. Butz, S. *Energy and Agriculture: Science, Environment, and Solutions*; Cengage Learning: Boston, MA, USA, 2014; ISBN 978-1-111-54108-8.
3. Ktenioudaki, A.; Butler, F.; Gallagher, E. The Effect of Different Mixing Processes on Dough Extensional Rheology and Baked Attributes. *J. Sci. Food Agric.* **2010**, *90*, 2098–2104. [CrossRef] [PubMed]
4. Hersey, J.A. Powder Mixing: Theory and Practice in Pharmacy. *Powder Technol.* **1976**, *15*, 149–153. [CrossRef]
5. Deveswaran, R.; Bharath, S.; Basavaraj, B.V.; Abraham, S.; Furtado, S.; Madhavan, V. Concepts and Techniques of Pharmaceutical Powder Mixing Process: A Current Update. Available online: https://www.semanticscholar.org/paper/Concepts-and-Techniques-of-Pharmaceutical-Powder-A-Deveswaran-Bharath/6c45c57bbfa769ea9a1f6c20b7decc7488958f61 (accessed on 9 April 2020).
6. Siti Zulaikha, R.; Sharifah Norkhadijah, S.I.; Praveena, S.M. Hazardous Ingredients in Cosmetics and Personal Care Products and Health Concern: A Review. *Public Health Res.* **2015**, *5*, 7–15.
7. Chen, S.J. In-Line, Continuous Mixing and Processing of Cosmetic Products. *J. Soc. Cosmet. Chem.* **1973**, *24*, 639–653.
8. Rielly, C.D. Mixing in food processing. In *Chemical Engineering for the Food Industry*; Food Engineering Series; Fryer, P.J., Pyle, D.L., Rielly, C.D., Eds.; Springer: Boston, MA, USA, 1997; pp. 383–433. ISBN 978-1-4615-3864-6.
9. Cuq, B.; Berthiaux, H.; Gatumel, C. Powder mixing in the production of food powders. In *Handbook of Food Powders*; Bhandari, B., Bansal, N., Zhang, M., Schuck, P., Eds.; Woodhead Publishing Series in Food Science, Technology and Nutrition; Woodhead Publishing: Shaxton, UK, 2013; pp. 200–229. ISBN 978-0-85709-513-8.
10. Géber, R.; Kocserha, I.; Orosz, V.; Simon, A.; Paróczai, C. Optimization of the Mixing Ratio of Two Different Clays Used for Ceramic Roof Tiles. *Mater. Sci. Forum* **2010**, *659*, 477–482. [CrossRef]
11. Zaspalis, V.T.; Kolenbrander, M. Mixing Homogeneity and Its Influence on the Manufacturing Process and Properties of Soft Magnetic Ceramics. *J. Mater. Process. Technol.* **2008**, *205*, 297–302. [CrossRef]
12. Ashurst, P.R. Carbonated Beverages. In *Reference Module in Food Science*; Elsevier: Amsterdam, The Netherlands, 2016; ISBN 978-0-08-100596-5.
13. Eh-Sheet, I.A. Improvement of Gasoline Octane Number by Blending Gasoline with Selective Components. Available online: https://www.researchgate.net/publication/311739917_Improvement_of_Gasoline_Octane_Number_by_Blending_Gasoline_with_Selective_Components (accessed on 9 April 2020).
14. Foong, T.M.; Morganti, K.J.; Brear, M.J.; da Silva, G.; Yang, Y.; Dryer, F.L. The Octane Numbers of Ethanol Blended with Gasoline and Its Surrogates. *Fuel* **2014**, *115*, 727–739. [CrossRef]
15. Kim, D.; Ates, N.; Kaplan, S.; Selbes, M.; Karanfil, T. Impact of Combining Chlorine Dioxide and Chlorine on DBP Formation in Simulated Indoor Swimming Pools. *J. Environ. Sci.* **2017**, *58*. [CrossRef]
16. Boswell, C.D.; Varley, J.; Boon, L.; Hewitt, C.J.; Nienow, A.W. Studies on the Impact of Mixing in Brewing Fermentation: Comparison of Methods of Effecting Enhanced Liquid Circulation. *Food Bioprod. Process.* **2003**, *81*, 33–39. [CrossRef]
17. Boaden, R.W.; Hutton, P.; Monk, C. A Computer-Controlled Anaesthetic Gas Mixer. *Anaesthesia* **1989**, *44*, 665–669. [CrossRef] [PubMed]
18. Das, S.; Chattopadhyay, S.; Bose, P. The Anaesthesia Gas Supply System. *Indian J. Anaesth.* **2013**, *57*, 489–499. [CrossRef]
19. Srivastava, U. Anaesthesia Gas Supply: Gas Cylinders. *Indian J. Anaesth.* **2013**, *57*, 500–506. [CrossRef] [PubMed]
20. Brookman, R.P.; Tibbott, B. Gases and Gas Equipment. In *Semiconductor Safety Handbook*; Bolmen, R.A., Ed.; William Andrew Publishing: Westwood, NJ, USA, 1998; pp. 472–580. ISBN 978-0-8155-1418-3.
21. Schlösser, M.; Seitz, H.; Rupp, S.; Herwig, P.; Alecu, C.; Sturm, M.; Bornschein, B. In-Line Calibration of Raman Systems for Analysis of Gas Mixtures of Hydrogen Isotopologues with Sub-Percent Accuracy. *Anal. Chem.* **2013**, *85*. [CrossRef]
22. Sari, S.; Timo, R.; Jussi, H.; Panu, H. Dynamic Calibration Method for Reactive Gases. *Meas. Sci. Technol.* **2019**, *31*, 034001. [CrossRef]
23. Rosenberg, E.; Hallama, R.A.; Grasserbauer, M. Development and Evaluation of a Calibration Gas Generator for the Analysis of Volatile Organic Compounds in Air Based on the Injection Method. *Fresenius J. Anal. Chem.* **2001**, *371*, 798–805. [CrossRef]

24. Debossan, L.; Carvalho, E.; Souza, M.; Gomes, J. Stability Study of Standards Used for Calibration of the Spectrophotometer Wavelength Scale. *J. Phys. Conf. Ser.* **2016**, *733*, 012026. [CrossRef]
25. Klink, T. Calibration and validation of spectrophotometers: A vendor's view. In *Analytical Spectroscopy Library*; Burgess, C., Jones, D.G., Eds.; Spectrophotometry, Luminescence and Colour; Science & Compliance; Elsevier: Amsterdam, The Netherlands, 1995; Volume 6, pp. 195–204.
26. Leidinger, M.; Schultealbert, C.; Neu, J.; Schütze, A.; Sauerwald, T. Characterization and Calibration of Gas Sensor Systems at Ppb Level—A Versatile Test Gas Generation System. *Meas. Sci. Technol.* **2017**, *29*, 015901. [CrossRef]
27. Hasenfratz, D.; Saukh, O.; Thiele, L. On-the-Fly Calibration of Low-Cost Gas Sensors. In *Proceedings of the Wireless Sensor Networks*; Picco, G.P., Heinzelman, W., Eds.; Springer: Berlin/Heidelberg, Germany, 2012; pp. 228–244.
28. Spinelle, L.; Gerboles, M.; Villani, M.G.; Aleixandre, M.; Bonavitacola, F. Field Calibration of a Cluster of Low-Cost Available Sensors for Air Quality Monitoring. Part A: Ozone and Nitrogen Dioxide. *Sens. Actuators B Chem.* **2015**, *215*, 249–257. [CrossRef]
29. Angely, L.; Levart, E.; Guiochon, G.; Peslerbe, G. Preparing Standard Samples for Detector Calibration in Gas Chromatographic Analysis of Gases. *Anal. Chem.* **1969**, *41*, 1446–1449. [CrossRef]
30. Köllö, Z.; Demange, D.; Bornschein, B.; Dörr, L.; Günther, K.; Kloppe, B. Calibrating a Gas Chromatograph to Measure Tritium Using Calorimetry. *Fusion Eng. Des.* **2009**, *84*, 1073–1075. [CrossRef]
31. Roberts, M.D. Calibration of the Pierre Auger Fluorescence Detector. *arXiv* **2003**, arXiv:astro-ph/0308410.
32. Lara-Ibeas, I.; Rodríguez-Cuevas, A.; Andrikopoulou, C.; Person, V.; Baldas, L.; Colin, S.; Le Calvé, S. Sub-Ppb Level Detection of BTEX Gaseous Mixtures with a Compact Prototype GC Equipped with a Preconcentration Unit. *Micromachines* **2019**, *10*, 187. [CrossRef]
33. Lara-lbeas, I.; Trocquet, C.; Nasreddine, R.; Andrikopoulou, C.; Person, V.; Cormerais, B.; Englaro, S.; Le Calvé, S. BTEX near Real-Time Monitoring in Two Primary Schools in La Rochelle, France. *Air Qual. Atmos. Health* **2018**, *11*, 1091–1107. [CrossRef]
34. Nasreddine, R.; Person, V.; Serra, C.A.; Schoemaecker, C.; Le Calvé, S. Portable Novel Micro-Device for BTEX Real-Time Monitoring: Assessment during a Field Campaign in a Low Consumption Energy Junior High School Classroom. *Atmos. Environ.* **2016**, *126*, 211–217. [CrossRef]
35. Nasreddine, R.; Person, V.; Serra, C.A.; Le Calvé, S. Development of a Novel Portable Miniaturized GC for near Real-Time Low Level Detection of BTEX. *Sens. Actuators B Chem.* **2016**, *224*, 159–169. [CrossRef]
36. Noël, F.; Serra, C.; Le Calvé, S. Design of a Novel Axial Gas Pulses Micromixer and Simulations of Its Mixing Abilities via Computational Fluid Dynamics. *Micromachines* **2019**, *10*, 205. [CrossRef] [PubMed]
37. Martin, N.A.; Goody, B.A.; Wang, J.; Milton, M.J.T. Accurate and Adjustable Calibration Gas Flow by Switching Permeation and Diffusion Devices. *Meas. Sci. Technol.* **2012**, *23*, 105005. [CrossRef]
38. Susaya, J.; Kim, K.-H.; Cho, J.; Parker, D. The Controlling Effect of Temperature in the Application of Permeation Tube Devices in Standard Gas Generation. *J. Chromatogr. A* **2012**, *1225*, 8–16. [CrossRef]
39. Lucci, P.; Pacetti, D.; Núñez, O.; Frega, N.G. *Current Trends in Sample Treatment Techniques for Environmental and Food Analysis*; IntechOpen: London, UK, 2012; pp. 127–164. ISBN 9799533079126. Available online: https://books.google.fr/books?id=uh6nzQEACAAJ (accessed on 8 June 2021).
40. Bhanot, D. Sample Injection Practices in Gas Chromatography. 2015. Available online: https://lab-training.com/2015/11/04/sample-injection-practices-in-gas-chromatography/ (accessed on 8 June 2021).
41. Dietz, W.A. Response Factors for Gas Chromatographic Analyses. *J. Chromatogr. Sci.* **1967**, *5*, 68–71. [CrossRef]

micromachines

MDPI

Article

Identifying Candidate Biomarkers of Ionizing Radiation in Human Pulmonary Microvascular Lumens Using Microfluidics—A Pilot Study

Larry J. Millet [1], Richard J. Giannone [2], Michael S. Greenwood [3], Carmen M. Foster [4], Kathleen M. O'Neil [5], Alexander D. Braatz [6] and Sandra M. Davern [7,*]

1. The Center for Environmental Biotechnology, The University of Tennessee, Knoxville, TN 37996, USA; lmillet1@utk.edu
2. Bioanalytical Mass Spectrometry Group, Oak Ridge National Laboratory, Oak Ridge, TN 37831-6229, USA; giannonerj@ornl.gov
3. Advanced Reactor Systems Group, Oak Ridge National Laboratory, Oak Ridge, TN 37831-6229, USA; greenwoodms@ornl.gov
4. Molecular and Cellular Imaging Group, Oak Ridge National Laboratory, Oak Ridge, TN 37831-6229, USA; fostercm1@ornl.gov
5. Oak Ridge Associated Universities, Oak Ridge, TN 37830, USA; oneilk@vt.edu
6. Nuclear Energy and Fuel Cycle Division, Oak Ridge National Laboratory, Oak Ridge, TN 37831-6229, USA; braatzad@ornl.gov
7. Radioisotope Science and Technology Division, Oak Ridge National Laboratory, Oak Ridge, TN 37831-6229, USA

* Correspondence: davernsm@ornl.gov

Abstract: The microvasculature system is critical for the delivery and removal of key nutrients and waste products and is significantly damaged by ionizing radiation. Single-cell capillaries and microvasculature structures are the primary cause of circulatory dysfunction, one that results in morbidities leading to progressive tissue and organ failure and premature death. Identifying tissue-specific biomarkers that are predictive of the extent of tissue and organ damage will aid in developing medical countermeasures for treating individuals exposed to ionizing radiation. In this pilot study, we developed and tested a 17 μL human-derived microvascular microfluidic lumen for identifying candidate biomarkers of ionizing radiation exposure. Through mass-spectrometry-based proteomics, we detected 35 proteins that may be candidate early biomarkers of ionizing radiation exposure. This pilot study demonstrates the feasibility of using humanized microfluidic and organ-on-a-chip systems for biomarker discovery studies. A more elaborate study of sufficient statistical power is needed to identify candidate biomarkers and test medical countermeasures of ionizing radiation.

Keywords: microfluidics; microvascular cells; γ radiation; proteomics; mass spectrometry

Citation: Millet, L.J.; Giannone, R.J.; Greenwood, M.S.; Foster, C.M.; O'Neil, K.M.; Braatz, A.D.; Davern, S.M. Identifying Candidate Biomarkers of Ionizing Radiation in Human Pulmonary Microvascular Lumens Using Microfluidics—A Pilot Study. *Micromachines* **2021**, *12*, 904. https://doi.org/10.3390/mi12080904

Academic Editor: Katsuo Kurabayashi

Received: 14 May 2021
Accepted: 21 July 2021
Published: 29 July 2021

1. Introduction

Insecurities inherent to the global proliferation of radioactive materials demand that innovative medical countermeasures be identified to improve the response to radiological release, exposure, and contamination. Medical countermeasures have the potential to ameliorate the negative health effects of acute radiation syndrome, the delayed effects of acute radiation exposure, and multiple organ dysfunction syndrome that could result from radiation exposure.

Longitudinal epidemiological studies of radiological accidents identify that "... the symptomatology of organ system involvement could be traced not only to the pathophysiology of the rapidly turning over cell renewal systems but—of equal or more importance—to the vascular system and specifically, to the endothelial components" [1]. Degeneration and failure of the vascular system are primary contributors to dysfunction and failure of the body's many organs, resulting in acute radiation syndrome, delayed effects of acute radiation exposure, and multiple organ dysfunction syndrome [2–4].

Chemical signals are produced and received by the single-cell-thick vascular endothelium that lines all blood vessels—signals that orchestrate blood vessel dilation, contraction, communication, and proliferative growth in tissues and organs of the body [5]. Following radiological insult, the vascular endothelium becomes the source and target of inflammation, which initiates a cascade of structural and functional breakdown of the vasculature. Capillary collapse and scarring result in a rigid ineffective circulatory system incapable of regeneration [6]. Nevertheless, a lack of mechanistic knowledge remains on the human vascular pathophysiology of radiation-induced injuries and an immune response to irradiated endothelial cells. Specifically, the identification of molecular pathways and targets is needed to develop medical countermeasure strategies that alleviate, stop, or reverse progressive vascular inflammation and fibrosis [7,8]. More concerted efforts are needed to identify the tissue-specific biomarkers that are predictive of the extent of tissue and organ damage and to identify medical countermeasures for treating individuals exposed to ionizing radiation [9,10]. The identification of biomarkers of radiation exposure has the potential to allow the resolving of pathobiological mechanisms of radiation-induced endothelial injury and permit the testing of effectiveness of candidate emergency medical countermeasures. With a greater knowledge of biomarkers of radiation exposure, early intervention may allow for natural cell recovery, or regeneration, to repair vascular damage [11–14]. However, the mechanisms of cellular response to gamma (γ) irradiation insults are not well understood, thus hampering efforts to design, select, or test candidate countermeasures.

Microfluidics enable the development and maintenance of a wide variety of organ-on-a-chip systems that constrain chemical volumes by limiting physical dimensions of cellular spaces in realms that are more in vivo-like than bulk dish or flask cultures [15–17]. Spatial-temporal chemical dynamics of cellular systems can easily be achieved at more realistic and relevant scales using microfluidics [18]. These miniature systems also minimize analyte dilution, thus allowing for improved detection of cell-released factors [19–23].

A vast number of microfluidic platforms have been created and implemented for vascular biology and bioengineering applications [24,25]. The bulk of these applications use co-culture constructs of multiple cell types toward the support of functional tissues [26–28]. Components involved in hydrogel formation, and fibroblast and microvascular endothelial cell seeding have been parameterized in microfluidics to instruct the development of prescribed microvascular network morphologies [29] for tissue specific engineering applications [30], including wound closure [31] and microtumor models [16,32], and for supporting the growth and maintenance of cellular spheroids [33].

Mass spectrometry of single cells has been achieved, but these methods are often destructive and only allow for a "snapshot" in time [34–36]. To our knowledge, no work to date demonstrates the ability to obtain a temporal resolution of early candidate biomarkers from direct radiation exposure of human cells or tissue in human organ or tissue systems using microfluidics.

In this work, we have established a 17 μL vascular lumen-on-a-chip platform for sustained cultures of human lung microvascular cells (HMVEC-L) in a monolayer under flow. This platform allows for the exposure of microvascular lumens to γ radiation for the temporal collection of cellular secretions through luminal perfusates and subsequent biomarker analysis. In this study, we generate proof-of-concept data using mass-spectrometry-based proteomics to analyze radiation damage. These results indicate that the detection of candidate biomarkers for radiation exposure can be obtained through this microfluidic platform. These microfluidics systems can be enhanced, augmented, and applied to this biological model or other human organ tissue structures to allow for further studies on irradiated cellular systems, neutrophil migration and immune modulation, γ-ionizing radiation induced fibrosis, and radiation-exposure metrics (phosphorylation of the histone protein H2AX (γ-H2AX)) of irradiated human-derived organs, in vitro.

2. Materials and Methods

2.1. Microfluidics

Chrome masks and microfluidics masters were produced in-house at the Center for Nanophase Material Sciences through an approved, peer-reviewed user project; microfluidic replicates were generated and assembled within the bio-affiliate laboratory at Oak Ridge National Laboratory. A laminar flow computational fluid dynamic model was performed with COMSOL Multiphysics 5.4 (Build: 346) utilizing the computer-aided design (CAD) import module for geometry uniformity between modeling and application. Single-layer microfluidics were fabricated from CAD drawings according to similar methods detailed previously [37]. Briefly, wafers were patterned with SU-8 2015 and developed with an SU-8 developer. Patterned wafers were etched with a modified Bosch process to produce raised structures (130 μm) for molding microfluidic channels. Wafers were coated with a trichloro(1H,1H,2H,2H-perfluoro-n-octyl)silane coating to facilitate replicate molding. Flexible attachment ports of silastic tubing (Helix Medical, Carpinteria, CA, USA) were molded in place during the polydimethylsiloxane (PDMS) pour-and-cure process using Duco cement (part #62435, Devcon, Hartford, CT, USA). PDMS replicas with integrated tubing were removed from the wafer, trimmed, cleaned (using 3M Scotch Magic tape, 3M, Maplewood, MN, USA), plasma treated, and baked to produce a single-layer PDMS microfluidic channel covalently attached onto a 25 mm × 75 mm microscope slide.

To prepare for cell culture studies, microfluidic systems were created, with care taken to prevent particulate contaminations. Microfluidics were sterilized under ultraviolet (UV) lights (2–3 min each side, Stratagene UV Stratalinker 2400). The channels were prepared for fluidic priming using a vacuum chamber to degas PDMS. The channels and glass were coated with poly-L-lysine (100 μg/mL water, sterile filtered), and rinsed with culture media to prevent the introduction of air before the direct seeding with HMVEC-L cells.

2.2. Cell Culture

HMVEC-L cells were procured (Lonza, Walkersville, MD, USA) and grown on poly-L-lysine coated microfluidics. Each microvascular microfluidic slide was retained in a 60 mm petri dish with sterile nanopure water to prevent dehydration during culture. T-75 flasks containing HMVEC-L cells were maintained in complete VascuLife media (Lifeline Cell Technologies, Frederick, MD, USA) with antibiotics included as supplements in VascuLife media kits. When the flask reached 80% confluence, cells were split using trypsin and VascuLife media. HMVEC-L cells were pelleted, rinsed, resuspended in 30 μL, perfused through the channels, and allowed to attach for 1 h before starting syringe pump for continuous perfusion at 0.5 μL/min. HMVEC-L cells in microfluidics were maintained up to 12 days in vitro. Immediately before γ irradiation, perfusion was stopped and the syringe with microfluidic system transported to the radiation source in a 37 °C, temperature-stable organ transport system (MT4-ET, B Medical Systems, Noblesville, IN, USA). After irradiation, cells were returned to the incubator and perfused with VascuLife media at 0.5 μL/min. Media were collected every 65 min to provide 30 μL media sample process volume for mass spectrometry; samples were immediately stored in microfuge tubes at −80 °C until they were processed for mass spectrometry. Stocks of untreated HMVEC-L cells were maintained in cryopreservation in VascuLife media supplemented with 10% fetal bovine serum and 10% dimethyl sulfoxide.

2.3. Gamma Irradiation Treatment

Confluent microvascular cultures (500–700 cells/mm^2) were transferred to the Cobalt-60 (Co-60) γ radiation source within 60 mm petri dishes held inside a thermostable organ transport chamber maintained at about 37 °C. Independent to microfluidic sample processing, bulk microvascular cell cultures were calibrated with exposure to Co-60 γ ray dosages of 1, 5, 10, and 20 Gy. For proof-of-concept in this work, we exposed cells to 10 Gy Co-60 γ ray for γ-H2AX labeling and proteomic measurements; 10 Gy is a center value in the range of irradiation exposures, as discussed below. During microfluidic sample exposure to γ rays or mock treatment (approximately 5–7 min), microfluidic cell systems were kept under similar room-temperature conditions; both treated and mock-treated devices were immediately placed in the prewarmed transport chamber and returned to the cell culture incubator for perfusion and media collection. Radiation damage was assessed by immunocytochemistry for the formation of double-stranded breaks in DNA, which produce phospho-histone H2A.X foci at serine residue 139 (γ-H2AX) foci. Immunocytochemistry of γ-H2AX following 10 Gy irradiated HMVEC-L in microfluidics was performed after 15 days in vitro.

2.4. Immunocytochemistry

Cells were fixed with 4% electron-microscopy-grade paraformaldehyde in phosphate-buffered saline (PBS) for 10–15 min, followed by a PBS rinse, then permeabilized with 0.25% Triton™ X-100 in PBS for 10 min. The blocking agent was MAXblock™ Blocking Medium (2 h, room temperature and overnight, 4 °C), followed by a 10 min wash in MAXwash™ washing medium (Active Motif Inc., Carlsbad, CA, USA). Anti-phospho-histone H2A.X (serine 139) primary antibody clone JBW301 (MilliporeSigma, Burlington, MA, USA) was diluted to 2 μg/mL. The secondary antibody, Anti-Mouse IgG (H+L), F(ab′)2 fragment CF™ 488A (MilliporeSigma) was diluted to 1 μg/mL. Both antibodies were diluted in MAXbind™ Staining Medium (Active Motif Inc.). The cells were incubated in 50 μL of primary antibody solution (2 h, room temperature). Three MAXwash rinses (5 min each) were performed after the primary antibody. Secondary antibody labeling (50 μL, 1 h, 37 °C) was followed by four washes with MAXwash washing medium (5 min each). Rhodamine phalloidin was used to label the actin cytoskeleton (30 min, room temperature, 0.0165 μM in PBS) (Biotium Inc., Fremont, CA, USA). Cells were triple rinsed before 4′,6-diamidino-2-phenylindole (DAPI) (0.02 μg/mL PBS) labeling (VWR International Ltd., Lutterworth, Leicestershire, England). For data acquisition, labeled cells were triple rinsed and covered with 20 μL of Mowiol 4–88 (MilliporeSigma) solution per the manufacturer's instructions (2008 Data sheet 777 Mowiol 4–88, Polysciences Inc., Warrington, PA, USA). The γ-H2AX imaging was performed with a Biotek Cytation-1 Cell Imaging Multi-Mode Reader (BioTek Instruments Inc., Winooski, VT, USA) and Gen5 Microplate Reader and Imager Software (catalog no. GEN5, BioTek Instruments Inc., Winooski, VT, USA).

2.5. Mass Spectrometry

HMVEC-L microvascular perfusates (30 μL) were stored at −80 °C until processed to obtain a biomarker library. The 30 μL samples of γ-irradiated and nonirradiated microvasculature microfluidics were prepared for liquid chromatography–tandem mass spectrometry (LC-MS/MS). Samples were diluted to 200 μL with denaturation/reduction buffer (4% sodium deoxycholate, 100 mM ammonium bicarbonate, 10 mM dithiothreitol), heated to 85 °C for 10 min, and treated with iodoacetamide (adjusted to 30 mM, 15 min at room temperature in the dark) to block disulfide-forming cysteine residues. Samples were passed through a 50 kDa molecular weight cut-off (MWCO) spin filter (Vivaspin 500; Sartorius, Gottingen, Germany) to remove serum albumin and other large proteins. Filter flow through was then transferred to a 10 kDa MWCO spin filter for on-filter proteolytic digestion with 1 μg proteomics-grade trypsin (Pierce) at 37 °C overnight, followed by a second addition of trypsin the following day for 3 h. The samples were then centrifuged at 12,500× *g* for 10 min to collect tryptic peptides that pass through the 10 kDa MWCO filter. The recovered peptides were then acidified to 0.5% formic acid, and the sodium deoxycholate precipitate removed with ethyl acetate as previously described [38]. Tryptic peptides were then autosampled onto an in-house-constructed, tri-phasic, 2D back column and desalted/cleaned up with a Vanquish uHPLC plumbed to a Q Exactive Plus high-resolution mass spectrometer (Thermo Fisher Scientific, Waltham, MA, USA) outfitted with a nanospray source as previously described [38]. Once desalted, peptides were then transferred via salt cut (500 mM ammonium acetate) to an in-house-pulled nanospray emitter (75 μm inner diameter fused silica packed with 30 cm of Kinetex C18 resin (5 micron particle size; Phenomenex, Torrance, CA, USA)) and analyzed by LC-MS/MS over a 120 min reversed-phase gradient as previously detailed [38].

2.6. Data Analysis

MS/MS data were then searched against the human proteome database (June 2017-build; UniProt) concatenated with common protein contaminants, and peptide spectral matches were scored, filtered, and assembled to proteins using the Proteome Discoverer software suite (Thermo Scientific; search algorithm: MS Amanda v.2; peptide scorer: IMP-Elutator) with the following parameters: MS1 tolerance $\leq$ 5 ppm, MS2 tolerance $\leq$ 0.02 Da, fully tryptic with 2 miscleavages, static modification on Cys = 57.0215 Da, dynamic modification on Met = 15.9949 Da. Peptide- and peptide-spectrum-match-level false discovery rates were assessed by hits to a reversed database and maintained at $\leq$0.01. High-confident peptide matches were then quantified by area under the curve and assembled to their respective proteins, whereby protein abundance was calculated by the summed abundance of all constituent peptides. Protein abundances were then log2-transformed and the resulting abundance distributions normalized and standardized by locally estimated scatterplot smoothing, median absolute deviation, and median centering via InfernoRDN [39]. Normalized protein abundances were then moved to Perseus proteome informatics software [40,41] to impute missing values and assess differences between irradiated and nonirradiated microvasculature microfluidic chips.

3. Results

The initial platform design consisted of 15 U-turn segments in a single serpentine channel containing one inlet and one outlet. Upon seeding cells in the system, a continuum of cell-seeding densities settling throughout the length of the microfluidic channel was repeatedly observed. Confluence was obtained within the first four U-turn segments within a day of cell seeding, whereas the remainder of the chip progressed from high confluence to patchy then sparse cell adherence. The result of the initial platform was that most of the chip contained an insufficient number of cells that prevented a confluent microvascular lumen from forming. The second and final experimental platform used one fewer U-turn segment (total 14), and an additional cell-loading port situated at the center of the platform (Figure 1). Minor changes were also made to channel corners and barrier walls of the outlet. Seven U-turn segments were situated on each side of the central loading channel between the cell seeding port and the seeding outlet ports; the resulting microfluidic device contained a single 17 μL channel.

A computational fluid dynamics model was used to provide a preliminary pressure loss and flow behavior estimate in the experimental platform used for cell culture, ionizing radiation exposure, and sample collection. Cell loading occurs with a clean dislodged pellet of cells, with approximately 50% to 60% cells per unit volume. We modeled pressure drops in these microfluidic platforms using prior data for media heavily loaded with cells; published values for the apparent viscosities of cell solutions indicate that a 58% cell volume fraction has a viscosity of 0.0122 Pa-s (12.2 centipoise), a 52% cell volume fraction has an apparent viscosity of 0.0097 Pa-s (9.7 centipoise), and a 34% cell volume fraction has a viscosity of 0.0026 Pa-s (2.6 centipoise) [42]. The boundary conditions of the simulation were held constant: the inlet at a fixed, uniform volumetric flowrate of 0.5 μL/min at a density of 1000.0 kg/m^3 and 1072.3 kg/m^3 (for a range of media densities) and a fixed outlet pressure of 1×10^5 Pa (atmospheric pressure, 1 ATM) [43]. For loading cells in the central portion of the channel, the modeling data (Figure 2) for the highest cell volume fraction (58%, viscosity 0.0122 Pa-s) show a negligible fluidic pressure drop of 0.002 Pa from the inlet to either outlet for bidirectional flow and 0.0004 Pa for the lowest cell volume fraction (34%, viscosity 0.0026 Pa) cited previously. The modeling data for loading cells in this platform through an end-to-end method show a pressure drop of 0.008 Pa for a 58% cell volume fraction (0.0122 Pa-s) and 0.0016 Pa for a 34% cell volume fraction (0.0026 Pa-s). Apart from channel corners, the modeling results also show that the flow velocity of cell loading through the central port is half (0.125–0.2 mm/s) that of the end-to-end loading (0.25–0.40 mm/s) (Figure 2) and that viscosity only affects pressure drop in these simulations, not flow velocity.

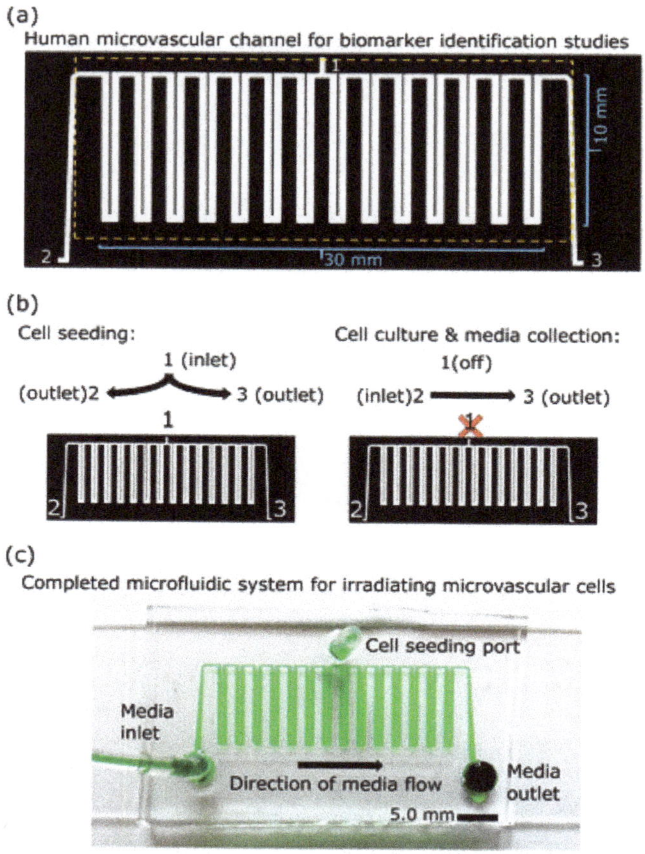

Figure 1. A large-scale microfluidic platform for generating microvascular lumens for biomarker analyses. (**a**) A schematic depicting the microfluidic platform design to maintain HMVEC-L cells and minimize dilution of analytes. The platform has three ports (1, 2, 3) that can be used for access points. Each of the 14 U-turns on the microfluidic chip contain 42 support posts (30 µm diameter) in channel dimensions of 21 mm (l), 550 µm (w), and 100 µm (h), giving 1.15 µL per U-turn; 14 U-turns and 15 interconnects (0.045 µL each) with a single inlet and outlet channel yields a total lumen volume of approximately 17 µL. Scale marks are shown (10 mm and 30 mm). (**b**) Diagrams for infusion flow direction for cell seeding and for culture maintenance are shown. Bilateral flow from center to end promotes uniform cell seeding throughout the long channel. During culture maintenance, the cell seeding port is sealed, and end-to-end perfusion flow maintains cell growth to achieve confluence of microvascular cells. (**c**) Image of a fabricated dye-filled microfluidic chip.

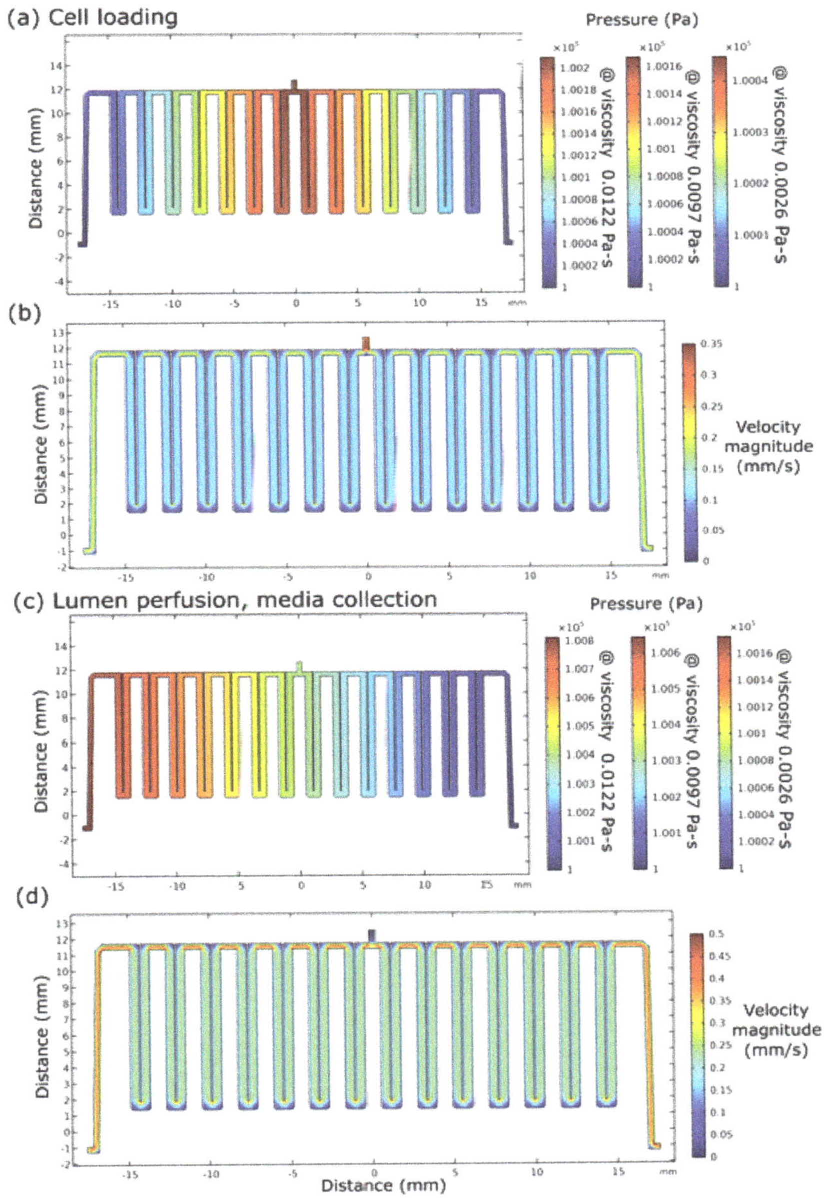

Figure 2. Computational fluid dynamics model velocity and pressure results in microfluidic channels for human microvascular endothelial lumens used for cell culture, irradiation, and media collection (**a**,**b**). During cell loading, a brief bidirectional flow introduces cells into the serpentine channel microenvironment under reduced velocity and pressure (**c**,**d**). End-to-end flow during lumen cell culture perfusion and media collection (0.5 µL/min) occurs with the middle cell loading port sealed; this perfusion system uses a slightly greater fluid velocity and pressure.

HMVEC-L cells were cultured to confluence for 6–8 days to produce a vascular lumen (Figure 3). Lumen perfusion with physiologically relevant flow rates (0.5 μL/min, 0.1–0.2 mm/s, 10 dyne/cm^2) maintained the HMVEC-L cells in microculture [44–46]. This was achieved based on successful perfusion values used in prior works and published values [47–49].

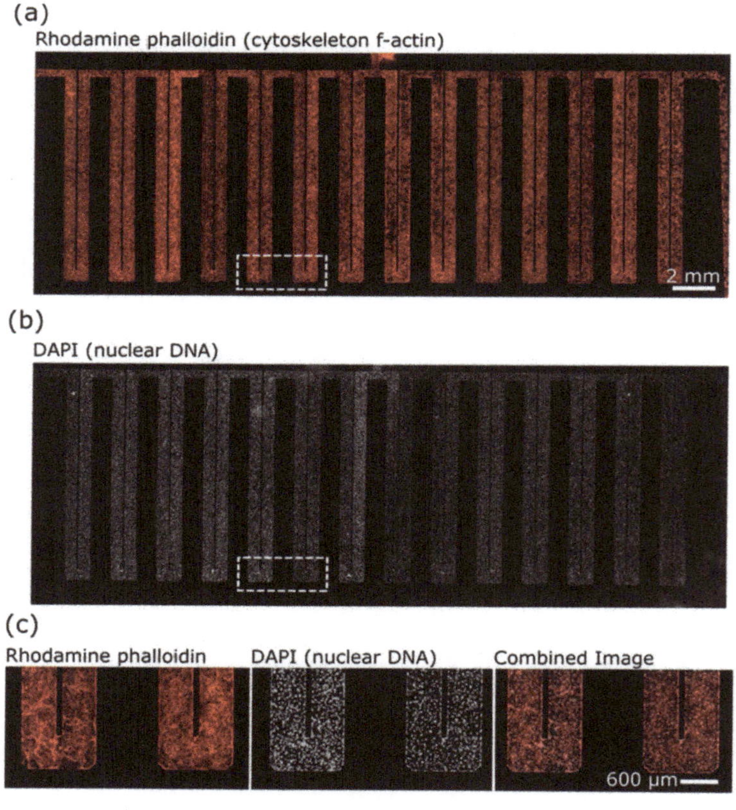

Figure 3. Human microvascular endothelial cells maintained in microfluidics for irradiation and biomarker identification. HMVEC-L cells were cultured with perfusion (0.25 to 0.5 μL/min) to confluence for 6–8 days to produce vascular lumens on a chip for irradiation and mass spectrometric analysis of perfusates. Composite images of the f-actin cytoskeleton, cellular nuclei, and merged images are shown here at 3 days in vitro. (**a**) Rhodamine-phalloidin-labeled f-actin cytoskeleton in HMVEC-L. (**b**) Cellular nuclei were labeled with DAPI. (**c**) Magnified end regions of microfluidic U-turn channels are identified by the dashed region of a–b.

Irradiation of human microvascular endothelial cells with γ rays induces double-stranded DNA breaks, damage that is evident in cells through immunocytochemistry for anti-phospho-histone H2A.X (serine 139) (γ-H2AX). Mock-treated cells (0.0 Gy, no γ irradiation) show few to no γ-H2AX foci per cell nuclei, whereas HMVEC-L cells exposed to 10 Gy of γ irradiation produce substantial γ-H2AX foci per nuclei at 3 h postexposure (Figure 4). The effect of ionizing radiation exposure and mock treatments to HMVEC-L cells was validated using γ-H2AX labeling and imaging.

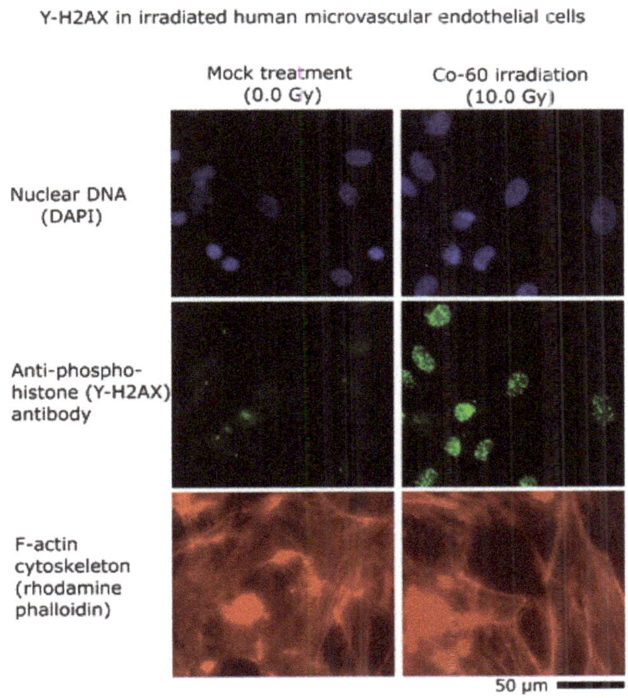

Figure 4. Irradiation of human microvascular endothelial cells with γ rays induces DNA damage. Images show DAPI-labeled cellular nuclei (top pair), γ-H2AX immunolabeling (middle pair), and rhodamine-phalloidin-labeled f-actin cytoskeleton (bottom pair). Double-stranded DNA breaks are evident in cells through immunocytochemistry for anti-phospho-histone H2A.X (serine 139), commonly termed γ-H2AX. Cells not exposed to γ rays (left image column) show very little γ-H2AX expression compared to cells exposed to 10 Gy of γ rays (right image column) from Co-60 source. The 50 µm scale bar applies to all six images.

Preliminary proteomic inquiry of a few samples (n = 3, irradiated; n = 2 nonirradiated) with temporally collected luminal flow through both irradiated and nonirradiated cells identified a total of 568 proteins in aggregate, 176 of which were found in all samples and conditions. Sixty-three proteins were exclusive to irradiated cells. To garner a more quantitative perspective, proteins were log2-transformed, normalized, and compared across conditions. High-abundance differences were assessed by t-test, and results were depicted as a volcano plot (Figure 5). Using a p-value cutoff of 0.05 (-log10 p-value = 1.301; y-axis) and a fold-change cutoff of one log2 unit (minimum 2× difference), 26 proteins were found to exhibit differential abundance, with 13 higher in irradiated cells and 13 higher in nonirradiated cells. Several interesting candidate proteins linked to cell growth and differentiation, cancer progression and prognosis, and other oncogenetic processes were found to be more abundant in irradiated cells. From these data, a preliminary summary of candidate protein biomarkers was obtained (Figure 5) and correlated to γ irradiation exposure. Through replication, randomization, and blinded sampling, we obtained 35 statistically significant candidate compounds that are different between γ irradiation exposure and mock controls. Protein abundance was derived at the peptide-level (MS1) by measuring chromatographic area under the curve and then normalizing. The results were compared across replicates (n = 3) and conditions with significant differences assessed by *t*-test and ANOVA using Fisher's least significant difference post hoc correction or Benjamini and Hochberg false discovery rate methods [50,51].

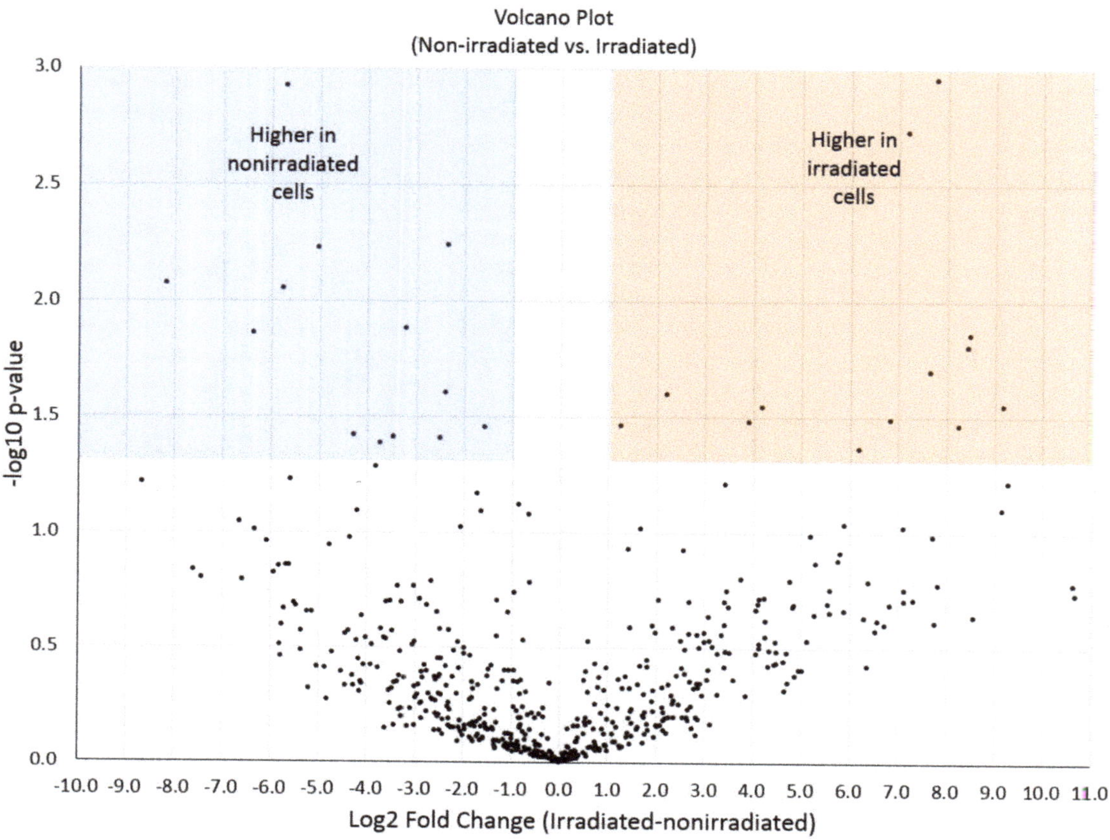

Figure 5. Volcano plot summarizing the proteomic results obtained from microvascular microfluidic lumen samples exposed to Co-60 γ radiation versus nonirradiated controls. Ultra-performance liquid chromatography (UPLC)-driven nanoelectrospray high-resolution tandem mass spectrometry and human proteome database analysis derived 26 statistically significant proteins that change in abundance between irradiated and nonirradiated tissue chip platforms (blue and orange boxed regions).

4. Discussion

Microfluidic technologies afford the ability to improve the chain of custody for cells and cellular solutions. Microfluidics combined with analytical chemistry have the potential to resolve the identities of cell-derived compounds needed for advancing biosystems research, producing medicinal compounds, and for advancing radiation biology [37,52,53]. The fluidic modeling data for this system were informative for guiding the perfusion of the microfluidic microvascular lumen. However, modeling does not fully capture the degree of challenge and complexity for loading cells in solution into a channel—essentially a flowing suspension of sticky semibuoyant spheres passing through a spatially restricted domain. Other researchers have modified the specific gravity of culture media formulations in an effort to adjust the maintenance frequency of sensitive cell populations [43]. The first channel built consisted of a long serpentine channel containing a single inlet and a single outlet. This two-port channel proved difficult for loading cells because they would adhere in an uneven distribution, progressing from high cell confluence to sparse seeding throughout the channel before clogging the fluidic channel. The resulting nonuniformity was unrecoverable, even if seeded from both directions, which became a barrier to throughput and platform success.

In the second version of the channel, we incorporated a loading port midway across the length of the channel, thus enabling facile loading of cells. After the cell loading and attachment, the port was sealed for end-to-end perfusion (Figure 1). Loading cells in microfluidics can be challenging. For example, cells can quickly attach one to another to form clusters that are too large for the channel and thus impede flow, a phenomenon not captured or described by these simulation data (Figure 2). Clumping and clogging can happen more readily in smaller-sized, low-volume channels, but the low cell density per unit volume can easily be overcome by cell division if the cells are mitotic and have low contact inhibition. As cells achieved and maintained confluence within our designed microfluidic device, the microvascular lumen they formed was then ready for irradiation, perfusion, and media collection.

The prior implementation of microfluidic systems for cell-based mass spectrometry has informed the process flow optimization for identifying analytes from cellular populations for advanced chemical analyses in peptidomics, metabolomics, and proteomics [21]. For example, microfluidic-assisted mass spectrometry has benefited neuroscience for the identification and label-free quantification of peptides released from neurons—fragile, intricate cells with high surface-to-area volume ratios [23,54,55]. Notwithstanding the successes amalgamating microfluidics with mass spectrometry, the pairing of these techniques is still in an optimization phase [22]. Microfluidics have been used for biodosimetry studies (quantitative nuclease protection assay) for high-throughput triage scenarios [56], γ-H2AX immunofluorescence cytometry [57,58], and gene expression assays of blood cells [59]. Recent reports indicate that more should be done to apply microfluidic radiobioassays to understand cellular physiology and pharmacokinetics [60].

In this pilot project, we chose to work with monocultures of primary human microvascular endothelial cells in microfluidics to obtain feasibility data for biomarker discovery studies of γ-irradiated cells. Using a 17 μL serpentine channel (Figures 1 and 2) enables adequate perfusion and serial collection of samples across time for mass spectrometry. Our use of a homotypic, confluent monolayer of cells structured in a vascular lumen is foundational for subsequent studies that use multiple cellular populations comprising more complex organ-on-a-chip tissue forms. Resolving chemical signatures of individual cell types is also a prerequisite for irradiation studies of heterogenous tissues, where the reconciliation of proteomic results may involve the deconvolving of unique, synergistic, tissue-derived identities. A tissue construct may provide more meaningful and unique signals that come from paracrine signaling between multiple cell types. Summating results from multiple homotypic samples may not provide the same level of proteomic complexity as would a heterogenous tissue. In this device, vascular cells arrive at confluence (Figure 3) before initiating γ ray exposure (Figure 4); they do so in a microfluidic volume with structure and operation to support serial sample collections for subsequent mass spectrometry. Our results show that this microfluidic domain has enough cellular content, media exposure, and process volume to provide sufficient sample quantity to identify early candidate biomarkers of radiation (Figure 5) in monolayers and in organ-on-a-chip tissue constructs. These results are encouraging for subsequent studies to identify early biomarkers of radiation injury.

To our knowledge, omics-level studies of human-derived cells or tissue structures using microfluidics and ionizing radiation have not been performed; rather, subject-limited applications with microfluidics have been achieved [53]. To this end, we performed this pilot feasibility study as a proof of principle for the microfluidic-based approach to radiobiological studies. Pilot studies should first confirm an innovation to be feasible and achieve or clarify certain design aspects but need not prove the effectiveness of the study. In this work, we have identified design rules and made modifications to procedural requirements that enable microfluidics to achieve a large enough microvascular lumen volume to enable reproducible mass spectrometry of the microvascular perfusate. This pilot study paves the way for a large-scale project of sufficient statistical power to enable further testing and validation—a study where a list of candidate biomarker identities that can be tested with medically relevant countermeasures are identified.

Two medical countermeasure drug application experiments have been performed by other researchers using human gut-on-a-chip and bone marrow-on-a-chip models [61,62]. In a separate study, global protein expression of neurovascular units and uncoupled blood brain barriers developed in organ-on-a-chip systems were analytically compared to identify novel expression changes and unpredicted identities of metabolites that maintain neuronal functionality [63]. Recently, we used microfluidics to achieve label-free, time-resolved exometabolite sampling of growing plant root exudates through nanoporous interfaces. The results showed that microfluidic sampling methods enabled tracking the distribution of sucrose-derivative products from the proximal and distal microenvironment and the exudate identification using extracted ion chromatography [19].

Our preliminary analyses focused on measuring differential abundances of secreted proteins across irradiated and nonirradiated cells. For this feasibility study, we chose to use 10 Gy irradiation dose as a midpoint between conventional lower dose radiotherapy (1–3 Gy) and higher dose radiosurgery (10 Gy to 35 Gy) [64,65]. High dose treatments are used in single dose fractions for large brain metastases and other focal lesions such as small peripheral lung tumors. Flash therapy administers higher doses of 10–20 Gy at a higher rate of delivery [66,67]. Therefore, evaluating the effect of higher radiation dose on biomarker release is important to consider for patients receiving radiotherapy. It is also an exposure to be considered for nuclear radiation events, such as a deliberate radiological release, e.g., a "dirty bomb".

Proteins of the microvasculature lumen were analyzed by high-resolution LC-MS/MS. The resulting peptide fragmentation data were searched and assigned peptides to their respective proteins, then contrasted across irradiated and nonirradiated microvasculature lumens to identify potential protein biomarkers of γ radiation exposure. Advances in bottom-up proteomic sample preparation methodologies enable microliter-scale proteomic measurements in minimal microliter-scale sample volumes [68]. Coupled with UPLC-driven nanoelectrospray high-resolution tandem mass spectrometry and microscale peptide fractionation, several hundreds to thousands of proteins can be reproducibly identified and quantified [69]. In this pilot study, we identified upward of 500 human proteins from the microvasculature flow through. Of these, 13 were found to be reproducibly increased in abundance upon ionizing radiation exposure. Encouragingly, several biomarker candidates related to cancer progression and/or prognosis have been found; however, pilot studies with low statistical power are not suitable for sequencing biomarkers. Experimental replicates greater than n = 3 are required before reporting statistically significant chemical identities. Nevertheless, these results are encouraging and lend credence to the applicability of microfluidic endothelial systems and organ-on-a-chip systems for biomarker discovery studies.

5. Conclusions

The few replicates of this pilot study demonstrate the feasibility of using humanized microfluidic and organ-on-a-chip systems for biomarker discovery studies. A more elaborate study of sufficient statistical power is needed to identify candidate biomarkers and test medical countermeasures of ionizing radiation to mitigate the resulting inflammatory and vascular disease. Obtaining these results with microfluidic systems is encouraging for further biomarker analysis studies and for implementing complex tissue architectures available through organ-on-a-chip systems. The primary benefit of these miniature systems is the ability to establish complex human tissue architecture for irradiation while minimizing analyte dilution. These microfluidic platforms easily enable spatiotemporal sampling—minutes to hours or even days—of the fluidic microenvironment off the same sample, thus eliminating the need for many multiples of parallel snapshots that are often required from bulk cultures. An added advantage of this approach is that conventional biochemical assays are moving to smaller analyte volumes through automated systems, thus allowing for improved detection of cell-released factors.

Author Contributions: Experiment design and manuscript writing, L.J.M., R.J.G., M.S.G., S.M.D.; experimental data collection and analysis assistance, C.M.F., K.M.D., A.D.B.; performed data analysis and managed the project, S.M.D. All authors have read and agreed to the published version of the manuscript.

Funding: This manuscript has been authored by UT-Battelle, LLC, under contract DE-AC05-00OR22725 with the US Department of Energy (DOE). The US government retains and the publisher, by accepting the article for publication, acknowledges that the US government retains a nonexclusive, paid-up, irrevocable, worldwide license to publish or reproduce the published form of this manuscript, or allow others to do so, for US government purposes. DOE will provide public access to these results of federally sponsored research in accordance with the DOE Public Access Plan (http://energy.gov/downloads/doe-public-access-plan accessed on 14 May 2021).

Acknowledgments: Notice: This work was funded by the Laboratory Directed Research and Development program of Oak Ridge National Laboratory, managed by UT-Battelle, LLC, for the US Department of Energy under contract DE-AC05-00OR22725. Microfluidic systems were fabricated and assembled through an approved user project of the Nanofabrication Research Laboratory at the Center for Nanophase Materials Sciences, which is a Department of Energy Office of Science User Facility. The content is solely the responsibility of the authors and does not necessarily represent the official views of the U.S. Department of Energy.

Conflicts of Interest: The authors declare no conflict of interest.

References

1. Fliedner, T.M.; Dörr, H.D.; Meineke, V. Multi-organ involvement as a pathogenetic principle of the radiation syndromes: A study involving 110 case histories documented in SEARCH and classified as the bases of haematopoietic indicators of effect. *Br. J. Radiol.* **2005**, *27*, 1–8. [CrossRef]
2. Akashi, M. Role of infection and bleeding in multiple organ involvement and failure. *Br. J. Radiol.* **2005**, *27*, 69–74. [CrossRef]
3. Kiang, J.G.; Olabisi, A.O. Radiation: A poly-traumatic hit leading to multi-organ injury. *Cell Biosci.* **2019**, *9*, 1–15. [CrossRef]
4. MacVittie, T.J.; Farese, A.M.; Parker, G.A.; Jackson, W. The Time Course of Radiation-induced Lung Injury in a Nonhuman Primate Model of Partial-body Irradiation with Minimal Bone Marrow Sparing: Clinical and Radiographic Evidence and the Effect of Neupogen Administration. *Health Phys.* **2019**, *116*, 366–382. [CrossRef] [PubMed]
5. Yannoutsos, A.; Levy, B.I.; Safar, M.E.; Slama, G.; Blacher, J. Pathophysiology of hypertension. *J. Hypertens.* **2014**, *32*, 216–224. [CrossRef] [PubMed]
6. Milliat, F.; Sabourin, J.-C.; Tarlet, G.; Holler, V.; Deutsch, E.; Buard, V.; Tamarat, R.; Atfi, A.; Benderitter, M.; François, A. Essential Role of Plasminogen Activator Inhibitor Type-1 in Radiation Enteropathy. *Am. J. Pathol.* **2008**, *172*, 691–701. [CrossRef] [PubMed]
7. Citrin, D.E.; Prasanna, P.G.S.; Walker, A.J.; Freeman, M.L.; Eke, I.; Barcellos-Hoff, M.H.; Arankalayil, M.J.; Cohen, E.P.; Wilkins, R.; Ahmed, M.M.; et al. Radiation-Induced Fibrosis: Mechanisms and Opportunities to Mitigate—Report of an NCI Workshop, 19 September 2016. *Radiat. Res.* **2017**, *188*, 1–20. [CrossRef] [PubMed]
8. Satyamitra, M.M.; Dicarlo, A.L.; Taliaferro, L. Understanding the Pathophysiology and Challenges of Development of Medical Countermeasures for Radiation-Induced Vascular/Endothelial Cell Injuries: Report of a NIAID Workshop, 20 August 2015. *Radiat. Res.* **2016**, *186*, 99–111. [CrossRef]
9. Singh, V.K.; Newman, V.L.; Romaine, P.L.; Hauer-Jensen, M.; Pollard, H.B. Use of biomarkers for assessing radiation injury and efficacy of countermeasures. *Expert Rev. Mol. Diagn.* **2016**, *16*, 65–81. [CrossRef] [PubMed]
10. Vicente, E.; Vujaskovic, Z.; Jackson, I.L. A Systematic Review of Metabolomic and Lipidomic Candidates for Biomarkers in Radiation Injury. *Metabolites* **2020**, *10*, 259. [CrossRef] [PubMed]
11. Shimura, T.; Yamaguchi, I.; Terada, H.; Okuda, K.; Svendsen, E.; Kunugita, N. Radiation occupational health interventions offered to radiation workers in response to the complex catastrophic disaster at the Fukushima Daiichi Nuclear Power Plant. *J. Radiat. Res.* **2015**, *56*, 413–421. [CrossRef] [PubMed]
12. Rafii, S.; Ginsberg, M.; Scandura, J.; Butler, J.M.; Ding, B.-S. Transplantation of Endothelial Cells to Mitigate Acute and Chronic Radiation Injury to Vital Organs. *Radiat. Res.* **2016**, *186*, 196–202. [CrossRef]
13. Fish, B.L.; Gao, F.; Narayanan, J.; Bergom, C.; Jacobs, E.R.; Cohen, E.P.; Moulder, J.E.; Orschell, C.M.; Medhora, M. Combined Hydration and Antibiotics with Lisinopril to Mitigate Acute and Delayed High-dose Radiation Injuries to Multiple Organs. *Health Phys.* **2016**, *111*, 410–419. [CrossRef] [PubMed]

14. Nepper-Christensen, S.; Heslet, L.; Bay, C. Acute radiation syndrome (ARS)—Treatment of the reduced host defense. *Int. J. Gen. Med.* **2012**, *5*, 105–115. [CrossRef] [PubMed]
15. Brown, T.D.; Nowak, M.; Bayles, A.V.; Prabhakarpandian, B.; Karande, P.; Lahann, J.; Helgeson, M.E.; Mitragotri, S. A microfluidic model of human brain (μHuB) for assessment of blood brain barrier. *Bioeng. Transl. Med.* **2019**, *4*, e10126. [CrossRef]
16. Sontheimer-Phelps, A.; Hassell, B.A.; Ingber, D.E. Modelling cancer in microfluidic human organs-on-chips. *Nat. Rev. Cancer* **2019**, *19*, 65–81. [CrossRef] [PubMed]
17. McLean, I.C.; Schwerdtfeger, L.A.; Tobet, S.A.; Henry, C.S. Powering ex vivo tissue models in microfluidic systems. *Lab Chip* **2018**, *18*, 1399–1410. [CrossRef]
18. Millet, L.; Gillette, M.U. New perspectives on neuronal development via microfluidic environments. *Trends Neurosci.* **2012**, *35*, 752–761. [CrossRef] [PubMed]
19. Patabadige, D.E.W.; Millet, L.; Aufrecht, J.; Shankles, P.G.; Standaert, R.F.; Retterer, S.T.; Doktycz, M.J. Label-free time- and space-resolved exometabolite sampling of growing plant roots through nanoporous interfaces. *Sci. Rep.* **2019**, *9*, 10272. [CrossRef] [PubMed]
20. Hu, Q.; Luni, C.; Elvassore, N. Microfluidics for secretome analysis under enhanced endogenous signaling. *Biochem. Biophys. Res. Commun.* **2018**, *497*, 480–484. [CrossRef] [PubMed]
21. Pedde, R.D.; Li, H.; Borchers, C.H.; Akbari, M. Microfluidic-Mass Spectrometry Interfaces for Translational Proteomics. *Trends Biotechnol.* **2017**, *35*, 954–970. [CrossRef] [PubMed]
22. Wang, X.; Yi, L.; Mukhitov, N.; Schrell, A.; Dhumpa, R.; Roper, M.G. Microfluidics-to-mass spectrometry: A review of coupling methods and applications. *J. Chromatogr. A* **2015**, *1382*, 98–116. [CrossRef]
23. Tharakan, R.; Tao, D.; Ubaida-Mohien, C.; Dinglasan, R.R.; Graham, D.R. Integrated Microfluidic Chip and Online SCX Separation Allows Untargeted Nanoscale Metabolomic and Peptidomic Profiling. *J. Proteome Res.* **2015**, *14*, 1621–1626. [CrossRef] [PubMed]
24. Costa, L.; Reis, R.L.; Silva-Correia, J.; Oliveira, J.M. Microfluidics for Angiogenesis Research. In *Advances in Experimental Medicine and Biology*; Springer Science and Business Media LLC: Cham, Switzerland, 2020; Volume 1230, pp. 97–119.
25. Haase, K.; Kamm, R.D. Advances in on-chip vascularization. *Regen. Med.* **2017**, *12*, 285–302. [CrossRef]
26. Moses, S.R.; Adorno, J.J.; Palmer, A.F.; Song, J.W. Vessel-on-a-chip models for studying microvascular physiology, transport, and function in vitro. *Am. J. Physiol. Physiol.* **2020**, *320*, C92–C105. [CrossRef]
27. Yang, F.; Cohen, R.N.; Brey, E.M. Optimization of Co-Culture Conditions for a Human Vascularized Adipose Tissue Model. *Bioengineering* **2020**, *7*, 114. [CrossRef]
28. Park, T.-E.; Mustafaoglu, N.; Herland, A.; Hasselkus, R.; Mannix, R.; Fitzgerald, E.A.; Prantil-Baun, R.; Watters, A.; Henry, O.; Benz, M.; et al. Hypoxia-enhanced Blood-Brain Barrier Chip recapitulates human barrier function and shuttling of drugs and antibodies. *Nat. Commun.* **2019**, *10*, 2621. [CrossRef]
29. Whisler, J.A.; Chen, M.B.; Kamm, R.D. Control of Perfusable Microvascular Network Morphology Using a Multiculture Microfluidic System. *Tissue Eng. Part C Methods* **2014**, *20*, 543–552. [CrossRef]
30. Song, H.G.; Lammers, A.; Sundaram, S.; Rubio, L.; Chen, A.X.; Li, L.; Eyckmans, J.; Bhatia, S.N.; Chen, C.S. Transient Support from Fibroblasts is Sufficient to Drive Functional Vascularization in Engineered Tissues. *Adv. Funct. Mater.* **2020**, *30*, 202003777. [CrossRef]
31. Tefft, J.B.; Chen, C.S.; Eyckmans, J. Reconstituting the dynamics of endothelial cells and fibroblasts in wound closure. *APL Bioeng.* **2021**, *5*, 016102. [CrossRef]
32. Wang, X.; Sun, Q.; Pei, J. Microfluidic-Based 3D Engineered Microvascular Networks and Their Applications in Vascularized Microtumor Models. *Micromachines* **2018**, *9*, 493. [CrossRef]
33. Nashimoto, Y.; Hayashi, T.; Kunita, I.; Nakamasu, A.; Torisawa, Y.-S.; Nakayama, M.; Takigawa-Imamura, H.; Kotera, H.; Nishiyama, K.; Miura, T.; et al. Integrating perfusable vascular networks with a three-dimensional tissue in a microfluidic device. *Integr. Biol.* **2017**, *9*, 506–518. [CrossRef]
34. Tsuyama, N.; Mizuno, H.; Masujima, T. Molecular and functional analysis of cellular phenomena using single-cell mass spectrometry. *Biol. Pharm. Bull.* **2012**, *35*, 1425–1431. [CrossRef] [PubMed]
35. Abouleila, Y.; Onidani, K.; Ali, A.; Shoji, H.; Kawai, T.; Lim, C.T.; Kumar, V.; Okaya, S.; Kato, K.; Hiyama, E.; et al. Live single cell mass spectrometry reveals cancer-specific metabolic profiles of circulating tumor cells. *Cancer Sci.* **2018**, *110*, 697–706. [CrossRef] [PubMed]
36. Zhang, L.; Vertes, A. Single-Cell Mass Spectrometry Approaches to Explore Cellular Heterogeneity. *Angew. Chem. Int. Ed.* **2018**, *57*, 4466–4477. [CrossRef]
37. Millet, L.J.; Lucheon, J.D.; Standaert, R.; Retterer, S.; Doktycz, M. Modular microfluidics for point-of-care protein purifications. *Lab Chip* **2015**, *15*, 1799–1811. [CrossRef]
38. Standaert, R.F.; Giannone, R.J.; Michener, J.K. Identification of parallel and divergent optimization solutions for homologous metabolic enzymes. *Metab. Eng. Commun.* **2018**, *6*, 56–62. [CrossRef]
39. Taverner, T.; Karpievitch, Y.V.; Polpitiya, A.D.; Brown, J.N.; Dabney, A.R.; Anderson, G.A.; Smith, R. DanteR: An extensible R-based tool for quantitative analysis of -omics data. *Bioinformatics* **2012**, *28*, 2404–2406. [CrossRef] [PubMed]
40. Tyanova, S.; Temu, T.; Sinitcyn, P.; Carlson, A.; Hein, M.Y.; Geiger, T.; Mann, M.; Cox, J. The Perseus computational platform for comprehensive analysis of (prote)omics data. *Nat. Methods* **2016**, *13*, 731–740. [CrossRef]

41. Tyanova, S.; Cox, J. Perseus: A Bioinformatics Platform for Integrative Analysis of Proteomics Data in Cancer Research. In *Methods in Molecular Biology*; Springer Science and Business Media LLC: Berlin, Germany, 2018; Volume 1711, pp. 133–148.
42. Zhong, J.-J.; Seki, T.; Kinoshita, S.-I.; Yoshida, T. Rheological characteristics of cell suspension and cell culture ofPerilla frutescens. *Biotechnol. Bioeng.* **1992**, *40*, 1256–1262. [CrossRef]
43. Kajiume, T.; Yuge, L.; Kawahara, Y.; Yoshimoto, R.; Sasaki, A.; Ide, T.; Asashima, M.; Kataoka, K.; Kobayashi, M. Floating culture promotes the maintenance of hematopoietic stem cells. *FEBS Lett.* **2007**, *581*, 4645–4650. [CrossRef] [PubMed]
44. Wu, D.; Birukov, K. Endothelial Cell Mechano-Metabolomic Coupling to Disease States in the Lung Microvasculature. *Front. Bioeng. Biotechnol.* **2019**, *7*, 172. [CrossRef] [PubMed]
45. Lipowsky, H.H. Shear Stress in the Circulation. In *Flow-Dependent Regulation of Vascular Function*; Springer Science and Business Media LLC: Berlin, Germany, 1995; pp. 28–45.
46. Paszkowiak, J.J.; Dardik, A.; Haven, N. *Basic Science Review Arterial Wall Shear Stress: Observations from the Bench to the Bedside*; Westminster Publications: Glen Head, NY, USA, 2003; Volume 37.
47. Jarvis, M.R.; Arnold, M.; Ott, J.; Krishnan, V.; Pant, K.; Prabhakarpandian, B.; Mitragotri, S. Detachment of ligands from nanoparticle surface under flow and endothelial cell contact: Assessment using microfluidic devices. *Bioeng. Transl. Med.* **2018**, *3*, 148–155. [CrossRef] [PubMed]
48 Liu, Z.; Mackay, S.; Gordon, D.M.; Anderson, J.; Haithcock, D.W.; Garson, C.J.; Tearney, G.J.; Solomon, G.M.; Pant, K.; Prabhakarpandian, B.; et al. Co-cultured microfluidic model of the airway optimized for microscopy and micro-optical coherence tomography imaging. *Biomed. Opt. Express* **2019**, *10*, 5414–5430. [CrossRef] [PubMed]
49. Tang, Y.; Soroush, F.; Sheffield, J.B.; Wang, B.; Prabhakarpandian, B.; Kiani, M.F. A Biomimetic Microfluidic Tumor Microenvironment Platform Mimicking the EPR Effect for Rapid Screening of Drug Delivery Systems. *Sci. Rep.* **2017**, *7*, 1–14. [CrossRef]
50. Naviaux, R.K.; Naviaux, J.C.; Li, K.; Bright, A.T.; Alaynick, W.A.; Wang, L.; Baxter, A.; Nathan, N.; Anderson, W.; Gordon, E. Metabolic features of chronic fatigue syndrome. *Proc. Natl. Acad. Sci. USA* **2016**, *113*, E5472–E5480. [CrossRef]
51. Noble, W.S.; MacCoss, M.J. Computational and Statistical Analysis of Protein Mass Spectrometry Data. *PLoS Comput. Biol.* **2012**, *8*, e1002296. [CrossRef] [PubMed]
52. Samiei, E.; Tabrizian, M.; Hoorfar, M. A review of digital microfluidics as portable platforms for lab-on a-chip applications. *Lab Chip* **2016**, *16*, 2376–2396. [CrossRef]
53. Lacombe, J.; Phillips, S.L.; Zenhausern, F. Microfluidics as a new tool in radiation biology. *Cancer Lett.* **2016**, *371*, 292–300. [CrossRef]
54. Croushore, C.A.; Supharoek, S.; Lee, C.Y.; Jakmunee, J.; Sweedler, J.V. Microfluidic Device for the Selective Chemical Stimulation of Neurons and Characterization of Peptide Release with Mass Spectrometry. *Anal. Chem.* **2012**, *84*, 9446–9452. [CrossRef]
55. Tillmaand, E.G.; Sweedler, J.V. Integrating mass spectrometry with microphysiological systems for improved neurochemical studies. *Microphysiol. Syst.* **2018**, *2*. [CrossRef] [PubMed]
56. Brengues, M.; Paap, B.; Bittner, M.; Amundson, S.; Seligmann, B.; Korn, R.; Lenigk, R.; Zenhausern, F. Biodosimetry on small blood volume using gene expression assay. *Health Phys.* **2010**, *98*, 179–185. [CrossRef] [PubMed]
57. Pope, I.; Barber, P.; Horn, S.; Ainsbury, E.; Rothkamm, K.; Vojnovic, B. A portable microfluidic fluorescence spectrometer device for γ-H2AX-based biological dosimetry. *Radiat. Meas.* **2011**, *46*, 907–911. [CrossRef]
58. Wang, J.; Song, W.; Song, Y.; Xu, D.; Zhang, M.; Pan, X.; Sun, Y.; Li, D. Quantitative evaluation of radiation dose by γ-H2AX on a microfluidic chip in a miniature fluorescence cytometer. *Radiat. Meas* **2014**, *62*, 71–77. [CrossRef]
59. Brengues, M.; Gu, J.; Zenhausern, F. Microfluidic module for blood cell separation for gene expression radiobiological assays. *Radiat. Prot. Dosim.* **2015**, *166*, 306–310. [CrossRef] [PubMed]
60. Liu, Z.; Lan, X. Microfluidic radiobioassays: A radiometric detection tool for understanding cellular physiology and pharmacokinetics. *Lab Chip* **2019**, *19*, 2315–2339. [CrossRef]
61. Jalili-Firoozinezhad, S.; Prantil-Baun, R.; Jiang, A.; Potla, R.; Mammoto, T.; Weaver, J.C.; Ferrante, T.C.; Kim, H.J.; Cabral, J.M.; Levy, O.; et al. Modeling radiation injury-induced cell death and countermeasure drug responses in a human Gut-on-a-Chip. *Cell Death Dis.* **2018**, *9*, 1–14. [CrossRef]
62. Torisawa, Y.-S.; Mammoto, T.; Jiang, E.; Jiang, A.; Mammoto, A.; Watters, A.L.; Bahinski, A.; Ingber, D.E. Modeling Hematopoiesis and Responses to Radiation Countermeasures in a Bone Marrow-on-a-Chip. *Tissue Eng. Part C Methods* **2016**, *22*, 509–515. [CrossRef] [PubMed]
63. Maoz, B.M.; Herland, A.; Fitzgerald, E.A.; Grevesse, T.; Vidoudez, C.; Pacheco, A.; Sheehy, S.P.; Park, T.-E.; Dauth, S.; Mannix, R.; et al. A linked organ-on-chip model of the human neurovascular unit reveals the metabolic coupling of endothelial and neuronal cells. *Nat. Biotechnol.* **2018**, *36*, 865–874. [CrossRef] [PubMed]
64. Xue, J.; Lacouture, T.; Grimm, J.; Goldman, H.W.; Ibbott, G.S.; Yorke, E.; Kubicek, G.J. Overview of dosimetric and biological perspectives on radiosurgery of multiple brain metastases in comparison with whole brain radiotherapy. *J. Radiosurg. SBRT* **2015**, *3*, 271–279.
65. Siva, S.; Bressel, M.; Kron, T.; Mai, T.; Le, H.; Montgomery, R.; Hardcastle, N.; Rezo, A.; Gill, S.; Higgs, B.; et al. Stereotactic Ablative Fractionated Radiotherapy versus Radiosurgery for Oligometastatic Neoplasia to the Lung: A Randomized Phase II Trial. *Int. J. Radiat. Oncol.* **2020**, *108*, S3–S4. [CrossRef]
66. Wilson, J.D.; Hammond, E.M.; Higgins, G.S.; Petersson, K. Ultra-High Dose Rate (FLASH) Radiotherapy: Silver Bullet or Fool's old? *Front. Oncol.* **2019**, *9*, 1563. [CrossRef] [PubMed]

67. Lin, B.; Gao, F.; Yang, Y.; Wu, D.; Zhang, Y.; Feng, G.; Dai, T.; Du, X. FLASH Radiotherapy: History and Future. *Front. Oncol.* **2021**, *11*. [CrossRef] [PubMed]
68. Lin, L.; Zheng, J.; Yu, Q.; Chen, W.; Xing, J.; Chen, C.; Tian, R. High throughput and accurate serum proteome profiling by integrated sample preparation technology and single-run data independent mass spectrometry analysis. *J. Proteom.* **2018**, *174*, 9–16. [CrossRef] [PubMed]
69. Lee, H.-J.; Kim, H.-J.; Liebler, D.C. Efficient Microscale Basic Reverse Phase Peptide Fractionation for Global and Targeted Proteomics. *J. Proteome Res.* **2016**, *15*, 2346–2354. [CrossRef] [PubMed]

micromachines

MDPI

Article

A Sensitivity-Enhanced Electrolyte-Gated Graphene Field-Effect Transistor Biosensor by Acoustic Tweezers

Yan Chen [1,2,†], Wenpeng Liu [3,*,†], Hao Zhang [3], Daihua Zhang [3] and Xiaoliang Guo [4,*]

1 Beijing Engineering Research Center of Industrial Spectrum Imaging, School of Automation and Electrical Engineering, University of Science and Technology Beijing, Beijing 100083, China; yanchen@ustb.edu.cn
2 School of Automation and Electrical Engineering, University of Science and Technology Beijing, Beijing 100083, China
3 College of Precision Instrument and Optoelectronics Engineering, Tianjin University, Tianjin 300072, China; haozhang@tju.edu.cn (H.Z.); dhzhang@tju.edu.cn (D.Z.)
4 College of Information Science and Technology, Beijing University of Chemical Technology, Beijing 100029, China
* Correspondence: liuwenpeng@tju.edu.cn (W.L.); gxl@mail.buct.edu.cn (X.G.)
† These authors contributed equally to this work.

Abstract: Low-abundance biomolecule detection is very crucial in many biological and medical applications. In this paper, we present a novel electrolyte-gated graphene field-effect transistor (EGFET) biosensor consisting of acoustic tweezers to increase the sensitivity. The acoustic tweezers are based on a high-frequency bulk acoustic resonator with thousands of MHz, which has excellent ability to concentrate nanoparticles. The operating principle of the acoustic tweezers to concentrate biomolecules is analyzed and verified by experiments. After the actuation of acoustic tweezers for 10 min, the IgG molecules are accumulated onto the graphene. The sensitivities of the EGFET biosensor with accumulation and without accumulation are compared. As a result, the sensitivity of the graphene-based biosensor is remarkably increased using SMR as the biomolecule concentrator. Since the device has advantages such as miniaturized size, low reagent consumption, high sensitivity, and rapid detection, we expect it to be readily applied to many biological and medical applications.

Keywords: electrolyte-gated graphene field-effect transistors; acoustic tweezers; solid mounted resonator (SMR)

Citation: Chen, Y.; Liu, W.; Zhang, H.; Zhang, D.; Guo, X. A Sensitivity-Enhanced Electrolyte-Gated Graphene Field-Effect Transistor Biosensor by Acoustic Tweezers. *Micromachines* **2021**, *12*, 1238. https://doi.org/10.3390/mi12101238

Academic Editor: Stefano Zampolli

Received: 19 September 2021
Accepted: 11 October 2021
Published: 13 October 2021

1. Introduction

Low-abundance biomolecule detection is very crucial in many biological and medical applications such as clinical diagnostics [1,2], drug discovery [3], and fundamental research [4]. As the conventional optical detection methods require professional skills and complex labeling processes, one of the trends of developing biosensors is exploring miniaturized analytical systems with reduced reagent consumption, high sensitivity, and rapid detection [5]. Various novel methods including surface plasmon resonance (SPR) [6], quartz crystal microbalance (QCM) [7], and electrochemical sensors [8], have been proposed. One of the novel methods is electrical biosensors that employ novel nanomaterials such as silicon nanowires and carbon nanotubes, which has attracted significant attention due to the advantages of its miniaturized size, low reagent consumption, high sensitivity, and rapid detection. Among various types of nanomaterials, graphene, which is a single layer two-dimensional crystal, has emerged as one of the most promising nanoplatforms [5]. Graphene shows extremely high mobility of $\sim 10^4$ cm^2 V^{-1} s^{-1} and large carrier capacities of $\sim 10^{12}$ cm^{-2}, even at room temperature without doping [9–15]. Meanwhile, as its electrical characteristics are sensitive to the surface conditions [11], the graphene can be used to design novel biosensors with high sensitivity [5,16,17].

Sensitivity is one of the key parameters of modern biosensors [18–20]. In this paper, we focus on how to further improve the sensitivity of the graphene-based biosensor. One of the fundamental factors hindering the improvement of the limit of detection is the mass transfer limitations [21,22]. To overcome the mass transfer limitations, various methods, including the electrokinetic-assisted method [23,24], magnetically-assisted method [25], and optically-assisted method [26–28], etc., have been proposed. The electrokinetic-assisted technique is limited due to the requirement of inherent charges of targets and the low ionic strength of solutions, thus hindering its applicability to many practical assays. The magnetically- and optically-assisted methods require extra labeling steps or complicated setups, which also limit their throughput.

In this work, we proposed a novel integrated biosensor consisting of an electrolyte-gated graphene field-effect transistor (EGFET) and solid mounted resonator (SMR) as acoustic tweezers. The SMR serves as an active biomolecule concentrator, which was verified in this paper. Thus, the sensitivity of the graphene-based biosensor was greatly enhanced. Meanwhile, the device fabrication is compatible with conventional semiconductor manufacturing processes and system-level integration, which means a low cost and miniaturization. Overall, the proposed method is very promising in biological and medical applications.

2. Methods

2.1. Mechanism Analysis

Acoustic tweezers are based on the interaction between acoustic waves and target particles within the fluids. One convenient way to generate acoustic waves is to use transducers made of piezoelectric materials. Piezoelectric materials can generate electrical polarization after applying mechanical stress, and the electrical polarization can also lead to mechanical deformation. Due to the piezoelectric effect, the radio frequency (RF) source applied to the SMR device can generate high-frequency longitudinal acoustic waves. When the piezoelectric material is immersed in solution, the high-frequency acoustic waves will trigger the acoustic streaming effects due to acoustic power leakage during the propagation of the acoustic waves into liquid [29–31]. High-frequency acoustic waves will contribute to the streaming flow strength due to the great energy attenuation coefficient after their transmission into liquid. Moreover, longitudinal waves can provide more effective energy coupling into the liquid than the shear mode acoustic waves to induce stronger turbulent flow [31]. To be exact, SMR devices have both features, namely high frequency and longitudinal acoustic waves, which can increase the acoustic streaming. Overall, the longitudinal wave and high frequency enable the SMR to be an effective actuator in triggering acoustic streaming and further driving biomolecules such as proteins and DNAs.

The fluid motions excited by the SMR device are visualized in Figure 1a. The longitudinal wave induces an upward fluid flow upon the top inner electrode (TIE) while the surrounding liquids are replenished, thus inducing a localized liquid flow. The biomolecules from the liquid around the device are subsequently transported to the graphene between the TIE and top outer electrode (TOE) by following the flow. Thus, an SMR-induced concentrating effect is generated around the gap between the TIE and TOE, as shown in Figure 1b.

The quantity of the biomolecules absorbed nonspecifically by the EGFET has a positive correlation with the accumulation of the biomolecules. The charges of the absorbed molecules can modulate the drain current of the EGFET; therefore, the Dirac point will drift. The accumulation of the biomolecules around the EGFET can be measured by monitoring the Dirac point. Since the SMR works as a biomolecule concentrator, the accumulation of the biomolecules around the EGFET is much higher than the bulk accumulation, thus lowering the detection limit of EGFET.

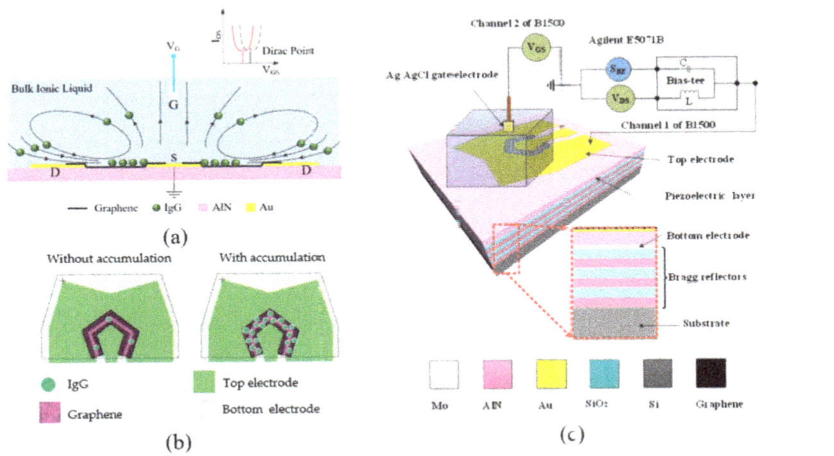

Figure 1. (**a**) The working principle of the EGFET with acoustic tweezers. (**b**) Schematic illustrating the accumulation of IgG molecules using SMR and the detection using EGFET. (**c**) The schematic diagram of the measurement setup.

2.2. Measurement Setup

After graphene patterning, the chip was mounted onto an evaluation board with the TIE and TOE connected to the center pin and outer shell of a coaxial small-A-type (SMA) connector. The board was then connected to a Bias-Tee to split RF and direct-current (DC) signals from each other. The configuration allows the dual modes to operate independently and simultaneously at distinct frequencies (GHz vs. DC) with negligible interference. The two output terminals of the Bias-Tee were connected to a network analyzer (Agilent E5071B, Agilent, Santa Clara, CA, USA) and the source meter (Agilent B1500, Agilent, Santa Clara, CA, USA) separately to characterize the RF and DC responses, respectively. The TOEs were grounded.

A 100 μL solution reservoir was mounted onto the device, and an Ag/AgCl reference electrode was used as the gate electrode to minimize environmental effects. The gate electrode was connected to the other channel of the Agilent B1500. The schematic diagram of the measurement setup is shown in Figure 1c.

IgG labeled with green fluorescence (Biosynthesis Biotechnology, Beijing, China) was used for the detection experiments. Various concentrations of IgG solution were obtained by stepwise diluting a certain volume of IgG solution (1 mM) into phosphate buffer saline (PBS). The IgG solution was added into the solution reservoir by pipette. The top-view images of the device were acquired by a microscope (Olympus BX53, Olympus, Tokyo, Japan) with a CCD camera (Olympus DP73, Olympus, Tokyo, Japan).

2.3. Fabrication of Device

The devices were fabricated on 100 mm undoped silicon wafers. The alternating aluminum nitride (AlN) and silicon dioxide (SiO_2) layers were deposited onto the silicon substrate to form the Bragg reflector [32], as shown in Figure 2a. The Bragg reflector transforms the impedance of the substrate to a near-zero or infinite value to avoid wave energy dissipation into the substrate [33]. The thicknesses of each AlN/SiO_2 pair were 1200/700, 1000/1300, and 1000/650 nm from bottom to top. The bottom electrode (BE) of the SMR was made of 600 nm thick molybdenum (Mo) on top of the Bragg reflector. The film was then patterned into isolated islands. A 1000 nm thick AlN film was then deposited on top of the BE. Orientation of the AlN crystal was along the c-axis. In the final step, the SMR was capped with a pair of gold (Au) top electrodes. The thicknesses of the Au electrodes and the chromium (Cr) adhesive layer were 300 and 50 nm, respectively. The

TIE was intentionally shaped like a pentagon to suppress spurious resonance in the device. The electrode area was configured to be 3.0×10^4 μm^2 so that the SMR has a characteristic impedance of 50 Ω to match the impedance of external circuits. The TOE was separated 15 μm away from the TIE. The area of TOE was several times larger than that of TIE. This geometrical arrangement ensures good electric field confinement within the active area under the TIE electrode. More details have been reported in the literature by us [32].

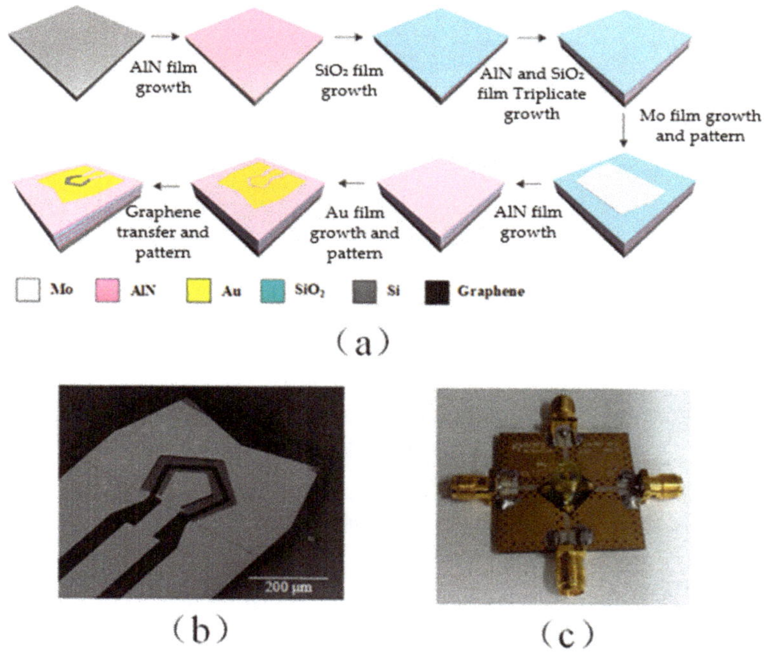

Figure 2. (**a**) The entire process flow SMR fabrication and graphene transfer. (**b**) SEM image of the device. The scale bar is 200 μm. (**c**) Image of a completed device mounted on the evaluation board.

Graphene field-effect transistors gated by electrolyte were used for detecting the biomolecules. Then, monolayer graphene film was transferred to the SMR. The chemical vapor deposition (CVD) grown monolayer graphene film (VIGON Technologies, Taipei, Taiwan) first went through a thorough RCA cleaning process to remove surface contaminants. Afterward, the device wafer was pretreated with O_2 plasma to clean the surface and condition it to a hydrophilic interface. The graphene film was then pressed against the wafer with the support of polymethylmethacrylate (PMMA) stamp and held in position for 2 h at room temperature for water evaporation. The fixture was then heated up to 150 °C with a hot plate and kept for 15 min. The heating step promotes van der Waals binding at the graphene–SMR interface, leaving an even and strongly-adhered graphene film after the removal of the PMMA stamp by immersing in acetone for 10 min, as shown in Figure 2b. The graphene film was trimmed using E-beam lithography, followed by O_2 plasma etching. We used a negative E-beam resist (Allresist, Zwickau, Germany), which introduced much fewer surface contaminants compared to most photoresists. Finally, the biosensor was mounted onto an evaluation board, as shown in Figure 2c.

2.4. Characterization of Device

The electrical performance of the device was measured from the RF port of the Bias-Tee, as shown in Figure 3a. The blue and red curves are the magnitude of electrical impedance at a frequency from 1.45 to 1.80 GHz when the device was in the air and immersed in liquid, respectively. The device behaved the same as a bare SMR without graphene in the air [29]. However, we saw a decrease in the Q factor from 275 to 69 when the device was immersed in liquid. The additional energy loss was primarily due to acoustic streaming effects in the liquid. The resonant frequency (1.645 GHz) was chosen to trigger the acoustic streaming effect due to its relatively large acoustic power leakage into the liquid [4]. The transfer characteristics of the EGFET on the SMR device in the liquid were also tested and are shown in Figure 3b. It behaved as a standard bio-polar EGFET. Surface doping of oxygen and moisture from ambient air leads to a p-type channel and shifts the charge neutrality point (CNP) toward positive gate bias of 0.17 V. Hole mobility of the device is derived to be ~87.4 $cm^2/(V{\cdot}s)$.

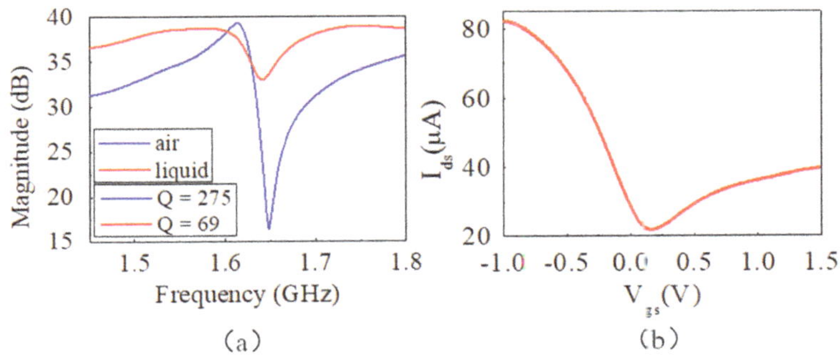

Figure 3. (**a**) Magnitude of electrical impedance at various frequencies and the Q values of the device in the air (blue curve) and in liquid (red curve). (**b**) Transfer characteristics of the graphene FET.

3. Results and Discussion

The solution containing green fluorescence-labeled IgG molecules was added into the reservoir. An RF signal of 1.645 GHz at a power of 10 mW was applied to the SMR device to generate acoustic waves. The IgG molecules were driven by acoustic waves to the gap between the TIE and TOE on the graphene. The green fluorescence intensity of the gap was observed to be increased on the gap through a fluorescence microscope, which indicated that the biomolecules were accumulated. After 10-min actuation, the fluorescent intensity of the biomolecules became stable. The final accumulated result is shown in Figure 4b. Compared with the other methods, such as the dielectrophoretic method [34] and ciliated micropillar-based method [35], which requires 30 min and 10 min, respectively, our method possesses an advantage with rapid accumulation.

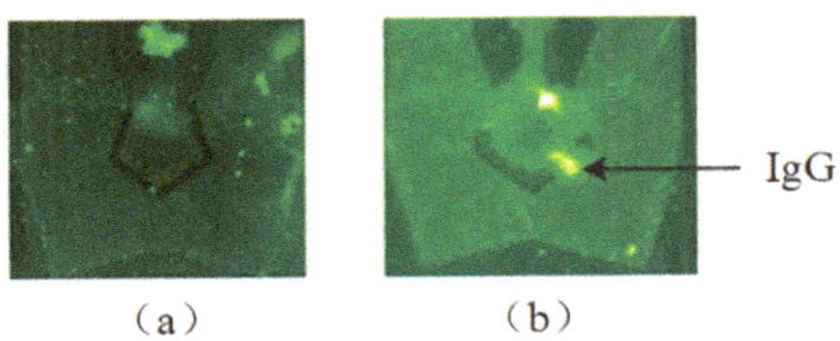

Figure 4. Fluorescence microscope images (**a**) without accumulation (**b**) with accumulation for 10 min.

After accumulation at various concentrations (1 nM, 5 nM, 10 nM, 25 nM, 50 nM, 75 nM, 100 nM, 125 nM, 150 nM, 175 nM, and 200 nM) of the IgG molecule by the SMR device, the Dirac points were calculated by sweeping the V_{gs}. Meanwhile, the Dirac points of the EGFETs merged into the same various concentrations of the IgG molecule without accumulation was also measured. Each concentration was measured three times. The Dirac point shifts against the concentration with and without accumulation are shown in Figure 5a. The detection limit of the sensor with accumulation is about 1 nM. Saturation concentrations are 100 nM. The Dirac point shift of the sensor without accumulation shows insignificant change in the concentration's range (1 nM~200 nM). The Dirac point shift of the sensor without accumulation is 0.37 V at 100,000 nM; this is close to the signal of the sensor without accumulation at 10 nM. We can see that the sensitivity of the device with accumulation is much higher than that without accumulation. The signal of the detection was increased by more than 4 orders. The data are fitted using the linear least-squares fitting method, as shown in Figure 5b. The sensor also has linearity with the R^2 of 0.96.

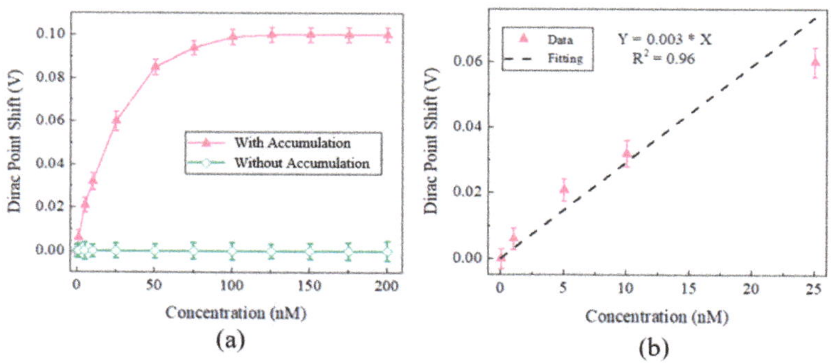

Figure 5. (**a**) Dirac point shifts at various concentrations with and without accumulation. (**b**) The relationship between the detection IgG concentration and the Dirac point shift.

4. Conclusions

In this paper, we have demonstrated a novel graphene-based transistor biosensor using acoustic tweezers to increase the sensitivity of the biosensor. Since the fabrication steps are compatible with IC processes, which means a small size, low reagent consumption, and low cost, this biosensor can be used as a disposable sensor to ensure repeatability and avoid cross contamination. The operating principle of the device has been described in detail and verified by experiment. We demonstrated that the biomolecules can be concentrated onto the graphene surface efficiently, which greatly increases the sensitivity. Overall, the device has great potential in biological and medical applications.

Author Contributions: Conceptualization, Y.C. and H.Z.; methodology, Y.C. and X.G.; software, Y.C.; validation, Y.C., W.L. and X.G.; formal analysis, Y.C.; investigation, Y.C.; resources, Y.C.; data curation, W.L.; writing—original draft preparation, X.G.; writing—review and editing, Y.C. and W.L.; visualization, Y.C.; supervision, H.Z. and D.Z.; project administration, Y.C.; funding acquisition, H.Z., Y.C., X.G. and D.Z. All authors have read and agreed to the published version of the manuscript.

Funding: This research was supported by the Fundamental Research Funds for the Central Universities (FRF-TP-19-032A1), National Natural Science Foundation (61904010), Natural Science Foundation of Beijing Municipality (4204109).

Conflicts of Interest: The authors declare no conflict of interest.

References

1. Mahshid, S.S.; Camire, S.; Ricci, F.; Vallee-Belisle, A. A Highly Selective Electrochemical DNA-Based Sensor That Employs Steric Hindrance Effects to Detect Proteins Directly in Whole Blood. *J. Am. Chem. Soc.* **2015**, *137*, 15596–15599. [CrossRef]
2. Kosaka, P.; Pini, V.; Ruz, J.J.; da Silva, R.; Ujue, M.; Ramos, D.; Calleja, M.; Tamayo, J. Detection of cancer biomarkers in serum using a hybrid mechanical and optoplasmonic nanosensor. *Nat. Nanotechnol.* **2014**, *9*, 1047–1053. [CrossRef]
3. Meister, S.; Plouffe, D.M.; Kuhen, K.L.; Bonamy, G.M.C.; Wu, T.; Barnes, S.W.; Bopp, S.E.; Borboa, R.; Bright, A.T.; Che, J.W.; et al. Imaging of Plasmodium Liver Stages to Drive Next-Generation Antimalarial Drug Discovery. *Science* **2011**, *334*, 1372–1377. [CrossRef]

4. Liu, W.P.; Pan, S.T.; Zhang, H.X.; Tang, Z.F.; Liang, J.; Wang, Y.Y.; Zhang, M.L.; Hu, X.D.; Pang, W.; Duan, X.X. A Universal Biomolecular Concentrator To Enhance Biomolecular Surface Binding Based on Acoustic NEMS Resonator. *ACS Cent. Sci.* **2018**, *4*, 899–908. [CrossRef]
5. Ohno, Y.; Maehashi, K.; Yamashiro, Y.; Matsumoto, K. Electrolyte-Gated Graphene Field-Effect Transistors for Detecting pH Protein Adsorption. *Nano Lett.* **2009**, *9*, 3318–3322. [CrossRef]
6. Scarano, S.; Mascini, M.; Turner, A.P.F.; Minunni, M. Surface plasmon resonance imaging for affinity-based biosensors. *Biosens. Bioelectron.* **2010**, *25*, 957–966. [CrossRef]
7. Mannelli, I.; Minunni, M.; Tombelli, S.; Mascini, M. Quartz crystal microbalance (QCM) affinity biosensor for genetically modified organisms (GMOs) detection. *Biosens. Bioelectron.* **2003**, *18*, 129–140. [CrossRef]
8. Guo, X.F.; Kulkarni, A.; Doepke, A.; Halsall, H.B.; Iyer, S.; Heineman, W.R. Carbohydrate-Based Label-Free Detection of Escherichia coli ORN 178 Using Electrochemical Impedance Spectroscopy. *Anal. Chem.* **2012**, *84*, 241–246. [CrossRef]
9. Novoselov, K.S.; Geim, A.K.; Morozov, S.V.; Jiang, D.; Zhang, Y.; Dubonos, S.V.; Grigorieva, I.V.; Firsov, A.A. Electric field effect in atomically thin carbon films. *Science* **2004**, *306*, 666–669. [CrossRef]
10. Novoselov, K.S.; Geim, A.K.; Morozov, S.V.; Jiang, D.; Katsnelson, M.I.; Grigorieva, I.V.; Dubonos, S.V.; Firsov, A.A. Two-dimensional gas of massless Dirac fermions in graphene. *Nature* **2005**, *438*, 197–200. [CrossRef]
11. Schedin, F.; Geim, A.K.; Morozov, S.V.; Hill, E.W.; Blake, P.; Katsnelson, M.I.; Novoselov, K.S. Detection of individual gas molecules adsorbed on graphene. *Nat. Mater.* **2007**, *6*, 652–655. [CrossRef]
12. Han, M.Y.; Ozyilmaz, B.; Zhang, Y.B.; Kim, P. Energy band-gap engineering of graphene nanoribbons. *Phys. Rev. Lett.* **2007**, *98*, 206805. [CrossRef]
13. Meric, I.; Han, M.Y.; Young, A.F.; Ozyilmaz, B.; Kim, P.; Shepard, K.L. Current saturation in zero-bandgap, topgated graphene field-effect transistors. *Nat. Nanotechnol.* **2008**, *3*, 654–659. [CrossRef]
14. Elias, D.C.; Nair, R.R.; Mohiuddin, T.M.G.; Morozov, S.V.; Blake, P.; Halsall, M.P.; Ferrari, A.C.; Boukhvalov, D.W.; Katsnelson, M.I.; Geim, A.K.; et al. Control of Graphene's Properties by Reversible Hydrogenation: Evidence for Graphane. *Science* **2009**, *323*, 610–613. [CrossRef]
15. Nagashio, K.; Nishimura, T.; Kita, K.; Toriumi, A. Mobility Variations in Mono- and Multi-Layer Graphene Films. *Appl. Phys. Express* **2009**, *2*, 025003. [CrossRef]
16. Dong, X.C.; Shi, Y.M.; Huang, W.; Chen, P.; Li, L.J. Electrical Detection of DNA Hybridization with Single-Base Specificity Using Transistors Based on CVD-Grown Graphene Sheets. *Adv. Mater.* **2010**, *22*, 1649–1653. [CrossRef]
17. Kakatkar, A.; Abhilash, T.S.; De Alba, R.; Parpia, J.M.; Craighead, H.G. Detection of DNA and poly-l-lysine using CVD graphene-channel FET biosensors. *Nanotechnology* **2015**, *26*, 125502. [CrossRef]
18. Lilja, H.; Ulmert, D.; Vickers, A.J. Prostate-specific antigen and prostate cancer: Prediction, detection and monitoring. *Nat. Rev. Cancer* **2008**, *8*, 268–278. [CrossRef]
19. Isaacs, J.S.; Xu, W.P.; Neckers, L. Heat shock protein 90 as a molecular target for cancer therapeutics. *Cancer Cell* **2003**, *3*, 213–217. [CrossRef]
20. Shangary, S.; Wang, S.M. Small-Molecule Inhibitors of the MDM2-p53 Protein-Protein Interaction to Reactivate p53 Function: A Novel Approach for Cancer Therapy. *Annu. Rev. Pharmacol.* **2009**, *49*, 223–241. [CrossRef]
21. Kim, J.; Junkin, M.; Kim, D.H.; Kwon, S.; Shin, Y.S.; Wong, P.K.; Gale, B.K. Applications, techniques, and microfluidic interfacing for nanoscale biosensing. *Microfluid. Nanofluid.* **2009**, *7*, 149–167. [CrossRef]
22. Kim, D.R.; Zheng, X.L. Numerical Characterization and Optimization of the Microfluidics for Nanowire Biosensors. *Nano Lett.* **2008**, *8*, 3233–3237. [CrossRef]
23. Liao, K.T.; Chou, C.F. Nanoscale Molecular Traps and Dams for Ultrafast Protein Enrichment in High-Conductivity Buffers. *J. Am. Chem. Soc.* **2012**, *134*, 8742–8745. [CrossRef]
24. Eijkel, J.C.T.; van den Berg, A. NANOFLUIDICS Tiny electrostatic traps. *Nature* **2010**, *467*, 666–667. [CrossRef]
25. Lien, K.Y.; Lee, W.C.; Lei, H.Y.; Lee, G.B. Integrated reverse transcription polymerase chain reaction systems for virus detection. *Biosens. Bioelectron.* **2007**, *22*, 1739–1748. [CrossRef]
26. Lee, S.; Hohng, S. An Optical Trap Combined with Three-Color FRET. *J. Am. Chem. Soc.* **2013**, *135*, 18260–18263. [CrossRef]
27. Grier, D.G. A revolution in optical manipulation. *Nature* **2003**, *424*, 810–816. [CrossRef]
28. Wang, K.; Schonbrun, E.; Steinvurzel, P.; Crozier, K.B. Trapping and rotating nanoparticles using a plasmonic nano-tweezer with an integrated heat sink. *Nat. Commun.* **2011**, *2*, 469. [CrossRef]
29. Nyborg, W.L. Acoustic Streaming Due to Attenuated Plane Waves. *J. Acoust. Soc. Am.* **1953**, *25*, 68–75. [CrossRef]
30. Mitome, H. The mechanism of generation of acoustic streaming. *Electron. Comm. Jpn.* **1998**, *81*, 1–8. [CrossRef]
31. Pan, S.T.; Zhang, H.X.; Liu, W.P.; Wang, Y.Y.; Pang, W.; Duan, X.X. Biofouling Removal and Protein Detection Using a Hypersonic Resonator. *ACS Sens.* **2017**, *2*, 1175–1183. [CrossRef]
32. Chen, Y.; Zhang, H.; Feng, Z.H.; Zhang, H.X.; Zhang, R.; Yu, Y.Y.; Tao, J.; Zhao, H.Y.; Guo, W.L.; Pang, W.; et al. Chemiresistive and Gravimetric Dual-Mode Gas Sensor toward Target Recognition and Differentiation. *ACS Appl. Mater. Interfaces* **2016**, *8*, 21742–21749. [CrossRef]
33. Chung, C.-J., Chen, Y.-C.; Cheng, C.-C.; Kao, K.-S. Fabrication and frequency response of solidly mounted resonators with 1/4λ mode configuration. *Thin Solid Films* **2008**, *516*, 5277–5281. [CrossRef]

34. Ibsen, S.D.; Wright, J.; Lewis, J.M.; Kim, S.; Ko, S.Y.; Ong, J.; Manouchehri, S.; Vyas, A.; Akers, J.; Chen, C.C.; et al. Rapid Isolation and Detection of Exosomes and Associated Biomarkers from Plasma. *ACS Nano* **2017**, *11*, 6641–6651. [CrossRef]
35. Wang, Z.; Wu, H.J.; Fine, D.; Schmulen, J.; Hu, Y.; Godin, B.; Zhang, J.X.; Liu, X. Ciliated micropillars for the microfluidic-based isolation of nanoscale lipid vesicles. *Lab Chip* **2013**, *13*, 2879–2882. [CrossRef]

micromachines

Communication

Colorimetric Sensing with Gold Nanoparticles on Electrowetting-Based Digital Microfluidics

Zhen Gu [1], Jing-Jing Luo [1], Le-Wei Ding [1], Bing-Yong Yan [1,*], Jia-Le Zhou [1], Jun-Gang Wang [2], Hui-Feng Wang [1,*] and Cong Kong [3,4,*]

1 Key Laboratory of Advanced Control and Optimization for Chemical Processes Ministry of Education, East China University of Science and Technology, Shanghai 200237, China; guzhen@ecust.edu.cn (Z.G.); ecust_ljj@163.com (J.-J.L.); 10182191@mail.ecust.edu.cn (L.-W.D.); zhou.jiale@ecust.edu.cn (J.-L.Z.)
2 School of Chemical and Environmental Engineering, Shanghai Institute of Technology, 100 Haiquan Road, Shanghai 201418, China; jgwang@sit.edu.cn
3 Shanghai Key Laboratory of Forensic Medicine, Academy of Forensic Science, Shanghai 200063, China
4 Key Laboratory of East China Sea Fishery Resources Exploitation, Ministry of Agriculture and Rural Affairs, East China Sea Fisheries Research Institute, Chinese Academy of Fishery Sciences, Shanghai 200090, China
* Correspondence: byyan@ecust.edu.cn (B.-Y.Y.); whuifeng@ecust.edu.cn (H.-F.W.); kongcong@gmail.com (C.K.)

Citation: Gu, Z.; Luo, J.-J.; Ding, L.-W.; Yan, B.-Y.; Zhou, J.-L.; Wang, J.-G.; Wang, H.-F.; Kong, C. Colorimetric Sensing with Gold Nanoparticles on Electrowetting-Based Digital Microfluidics. *Micromachines* **2021**, *12*, 1423. https://doi.org/10.3390/mi12111423

Academic Editor: Stefano Zampolli

Received: 31 October 2021
Accepted: 16 November 2021
Published: 19 November 2021

Publisher's Note: MDPI stays neutral with regard to jurisdictional claims in published maps and institutional affiliations.

Abstract: Digital microfluidic (DMF) has been a unique tool for manipulating micro-droplets with high flexibility and accuracy. To extend the application of DMF for automatic and in-site detection, it is promising to introduce colorimetric sensing based on gold nanoparticles (AuNPs), which have advantages including high sensitivity, label-free, biocompatibility, and easy surface modification. However, there is still a lack of studies for investigating the movement and stability of AuNPs for in-site detection on the electrowetting-based digital microfluidics. Herein, to demonstrate the ability of DMF for colorimetric sensing with AuNPs, we investigated the electrowetting property of the AuNPs droplets on the hydrophobic interface of the DMF chip and examined the stability of the AuNPs on DMF as well as the influence of evaporation to the colorimetric sensing. As a result, we found that the electrowetting of AuNPs fits to a modified Young–Lippmann equation, which suggests that a higher voltage is required to actuate AuNPs droplets compared with actuating water droplets. Moreover, the stability of AuNPs was maintained during the processing of electrowetting. We also proved that the evaporation of droplets has a limited influence on the detections that last several minutes. Finally, a model experiment for the detection of Hg^{2+} was carried out with similar results to the detections in bulk solution. The proposed method can be further extended to a wide range of AuNPs-based detection for label-free, automatic, and low-cost detection of small molecules, biomarkers, and metal ions.

Keywords: digital microfluidic; gold nanoparticle; electrowetting; colorimetric sensing; analytical instrument

1. Introduction

In the last decades, digital microfluidics (DMF) have attracted widespread attention as a new lab-on-a-chip approach for programmable manipulating of droplets at the volume of nano-to-micro liters [1,2]. The droplets on the DMF chip can be actuated in parallel to mix, separate, and dispose, which can be used in wide applications such as chemical reactions, automatic detection, and cell culture [3–5]. DMF based on electrowetting on dielectric (EWOD) is promising to be used in portable applications such as deployable environmental monitoring and point-of-care testing (POCT) [6], since it drives the droplets only requiring configuration of the voltage applied on an electrode array, which has advantages including accurate volume control, small size, low reagent cost, and low power consumption [7–10]. The key to actuating the droplet on DMF is to fabricate a dielectric layer on the electrode array with features such as thin, insulated, and hydrophobic [11].

Then, a lateral force on the droplets will be generated by applying a biased voltage between the electrodes. The direction and velocity of droplet movements can be programmed according to the detection process even after the fabrication of the EWOD-based DMF chips. As various methods have been developed to fabricate DMF chips based on substrates such as silicon wafer, ITO glass, PCB board, and flexible polymers, the DMF technique is able to combine with a variety of analytical methods such as electrochemiluminescence [12], electrochemistry [13], and colorimetric sensing [14]. Currently, the EWOD-based DMF has been widely used in biosensing, environmental monitoring, and food safety, such as single-cell sequencing [2,15], cell invasion monitoring [5], detection of growth factors in embryo culture medium [9], bacterial classification [16], marine pollution monitoring [17], and detection of food nitrite [14].

Nanomaterials have been widely used in analytical methods as sensors or indicators. Among these, gold nanoparticles (AuNPs) with unique optical and chemical properties are widely used for the detection of small molecules, heavy metals, and biological molecules, including DNA, RNA, and proteins [18]. The colorimetric assay based on AuNPs has the advantage of low limit of detection (LOD), because the interaction between the surface electrons on the AuNPs and the incident light causes the effect of localized surface plasmon resonance (LSPR), which highly amplify the light scattering of AuNPs [19]. Usually, colorimetric techniques based on AuNPs is realized by modulating the distances between AuNPs with target molecules. For example, the trans-cleavage activity of CRISPR/Cas12a is used to facilitate the assembly of AuNPs for the detection of SARS-CoV-2 [20]. Antibiotics are detected based on its adsorption on the AuNPs, which prevents the color changes induced by the aggregation of AuNPs by the addition of salt solution [21]. Therefore, the AuNPs is promising to be applied as sensors for automatic detections based on the DMFs. Although previous studies have demonstrated the possibility to control nanosuspensions by using electrowetting [22], the electrowetting property of AuNPs and the effect of solvent evaporation to the results of colorimetric sensing on DMFs were not concerned.

With these insights, in this study, we developed a novel DMF system for the manipulation of AuNPs droplets based on electrowetting and examined the ability of AuNPs for colorimetric sensing on the DMFs. We first measured the mobility of AuNPs droplets on the DMF chip. Then, the stability of AuNPs droplets during electrowetting was also investigated by monitoring the droplet absorbance during actuating on the DMF chip. In addition, the evaporation effect to the colorimetric sensing was also evaluated. As a module experiment, the detection of Hg^{2+} based on AuNPs was automatically conducted by using the proposed DMFs with integrated two-wavelength measuring module, and the results were consistent with the detections performed in a bulk solution, which suggests that the proposed method could be further developed for a versatile analytical platform for portable applications, such as point-of-care testing, in-field environment monitoring, and food safety.

2. Materials and Methods

2.1. Materials and Reagents

Hg^{2+} standard solution (100 mg/L) was purchased from Inorganic Ventures, Inc., Christiansburg, VA, USA. Trisodium citrate, lysine, $HAuCl_4$ was purchased from Sinopharm Chemical Reagent Co., Ltd. All buffers and solutions were made with deionized water purified by Milli-Q (Millipore, Billerica, MA, USA).

2.2. Setup of DMF Platform

An integrated DMF system was used for the automatic manipulation and sensing of the droplets as shown in Figure 1, which contains a two-plate DMF chip, optical sensor (TSL2571, ams AG, Graz, Austria), LED light source (520 nm, XLamp, Cree Inc., Durham United States), and control board with a microcontroller [23]. The digital camera is vertically mounted above the DMF chip. As shown in Figure 1a, the bottom plate of the DMF chip was made of printed circuit board techniques (FASTPCB Technology, Shenzhen China), which comprised 58 actuation electrodes (2.25 mm × 2.25 mm) and 3 reservoir electrodes (9.72 mm × 6.95 mm) with 0.1 mm gaps between each. The edges of the electrodes were designed with cross-fingers to promote the movement of droplets (Figure 1c–e). A PTFE film (10 μm, Hongfu Material, Dongguan, China) tightly attached on the electrode region of the bottom plate performs as both dielectric and hydrophobic layers. The top plate was made of indium tin oxide (ITO) glass with a spin coating of CYTOP (CTX-809SP2, AGC Inc., Tokyo, Japan) for surface hydrophobization. To increase the sensitivity of colorimetric sensing, the DMF chip was assembled by stacking the two plates with a gap of 0.9 mm by using double-sided tape as spacers. The control board was used for executing the movement of droplets, acquiring data from the optical sensor and communicating with computer software. A 3D-printed case was designed for the whole system to reduce the influence of environmental light on optical sensing.

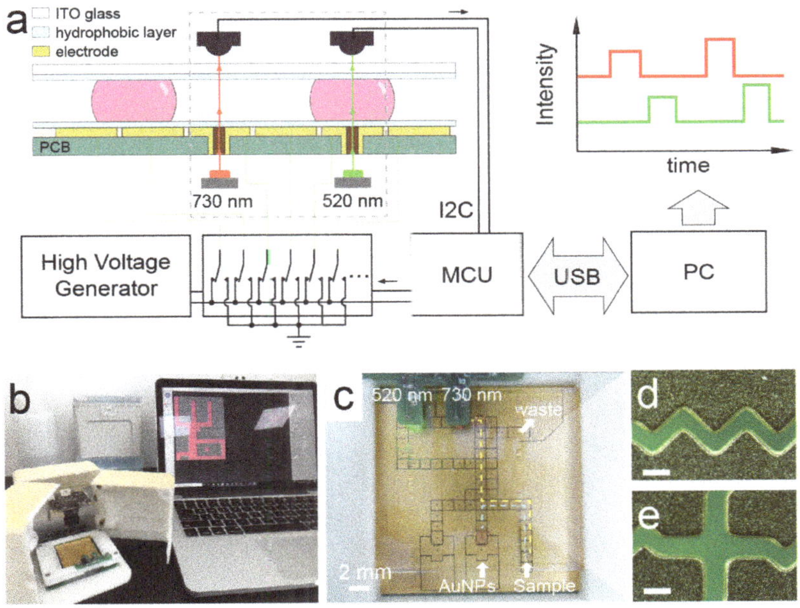

Figure 1. (**a**) Scheme of the digital microfluidic system for colorimetric detection at two wavelengths (730 nm and 520 nm). (**b**) Photograph of the digital microfluidic system. (**c**) Photograph of the digital microfluidic chip integrated with optical sensing elements. (**d**,**e**) are the microscopy images of the cross-fingers of the electrodes and the gaps between the electrodes. The scale bar equals 100 μm.

2.3. Synthesis and Characterization of AuNPs

AuNPs were prepared by the reduction of $HAuCl_4$ with trisodium citrate according to the reported procedure as follows [18]: 0.1% trisodium citrate was rapidly added to a boiling aqueous solution of 0.01% $HAuCl_4$. After the solution was heated for 20 min with vigorous stirring, the wine red AuNPs were obtained. Then, the products were cooled at room temperature. The average particle size of the AuNPs was determined to be 23.1 ± 4.5 nm from SEM. The concentration of the AuNPs (0.25 nM) was evaluated based on the size and its absorbance at 450 nm measured by a UV/Vis spectrometer [24]. The contact angle of the AuNPs droplets controlled by EWOD was measured through a contact angle meter.

2.4. Absorbance Measurement of Droplets

To measure the absorbance of the droplets at both 520 nm and 730 nm on the DMF, two apertures were fabricated in the sensing electrodes as light path (Figure 1). A two-wavelength measuring module with two LEDs (520 nm/730 nm) and two light-to-digital converters were integrated in the system. The LEDs were used as a light source and mounted below the apertures, respectively. Light-to-digital converters (TSL2571, ams AG, Graz, Austria) were mounted above the apertures for continuously acquiring the light intensity and transmitting the data to a microcontroller by using the I2C protocol. The absorbance of AuNPs was measured when the droplets moved to the sensing electrode.

3. Results and Discussion

3.1. Manipulation of the AuNPs Droplets on DMF

For EWOD-based DMF, the lateral force to drive the droplets comes from the changing of contact angle by applying an electric field, which usually follows the Young–Lippmann equation (Equation (1)). As the AuNPs are dispersed in aqueous solution, it is reasonable to be controlled by the electrowetting DMF. Meanwhile, the interactions between nanoparticles and the solvent under high electric field may not be ignored. To evaluate the significance of the electrowetting effect on the AuNPs droplets, the changing of the contact angle of both the AuNPs droplets and water droplets with a volume of 10 μL on the DMF chip was measured by applying a DC voltage ranging from 0 to 300 V. Here, AuNPs with a diameter of ≈25 nm was studied, as it is widely used in colorimetric assays. As a result, the contact angles of both the droplets were almost the same (≈105°) without applied voltage (Figure 2a). With the increase in applied voltage, the contact angle of water droplets appears to be lower than that of AuNPs droplets before the saturation of contact angle. The results of the water droplets before contact angle saturation were fitted to the Young–Lippmann equation (R^2 = 0.92). On the contrary, the results of the AuNPs droplets were not fitted to the Young–Lippmann equation, which states that the EWOD force is not proportional to the square of the applied voltage. The difference in the electrowetting contact angle between the AuNPs droplets and the water droplets may result from the change in interfacial energy induced by the interaction of nanoparticles at the solid–liquid interface of the AuNPs droplets [22,25]. Therefore, a modified form of the Young–Lippmann equation for nanosuspensions by adding an item representing the change of the solid–liquid surface tension was used to fit the electrowetting contact angle of AuNPs droplets [22]. As shown in Equation (2), the effect of the AuNPs increases linearly with the applied voltage, where A is a constant, and χ is the effect of nanoparticle concentration/volume in one lateral dimension. The fitting result was agreed well with the experimental data (R^2 = 0.99), supporting that the modified form of the Young–Lippmann equation was suitable to explain the electrowetting of the AuNPs droplets. Therefore, the result suggested that a higher applied voltage is needed to actuate the AuNPs droplet than for the water droplet to ensure the success of the manipulation, which should be taken into consideration in practice.

$$\cos\theta = \cos\theta_0 + \frac{\varepsilon_r\varepsilon_0}{2\gamma_{LG}d}V^2 \quad (1)$$

$$\cos\theta = \cos\theta_0 + \frac{\varepsilon_r\varepsilon_0}{2\gamma_{LG}d}V^2 + A\chi^{1/3}V \quad (2)$$

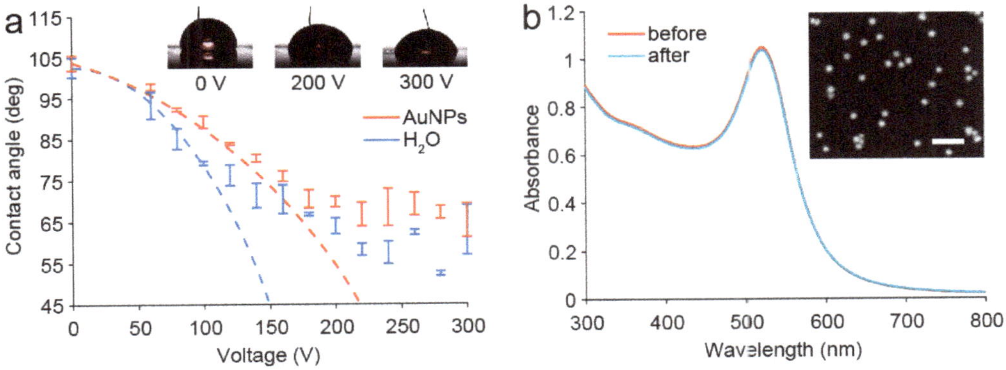

Figure 2. (**a**) Contact angles of the AuNPs droplets and water droplets under the applied voltage. Each data point is the mean of three measurements and error bars show standard deviation. (**b**) UV/Vis absorbance spectra of the AuNPs before and after the experiment and the SEM image of the synthesized AuNPs (inset, the scale bar equals 100 nm).

To examine the stability of the AuNPs actuated by the DMF, the AuNPs droplets were manipulated to be moving for 5 min (≈600 mm moving distance), and its UV/Vis absorption spectra were measured before and after the movement (Figure 2b). The results indicated that there was no aggregation or sediment in the AuNPs droplets as no additional peak was observed in the UV/Vis absorption spectra and the absorbance spectra of the AuNPs was nearly equal before and after the experiment. This reveals that the AuNPs is stable to be controlled by the DMF. Therefore, it could be further used in applications for DMF-based automatic colorimetric detections.

3.2. Evaporation Effect during the Colorimetric Sensing of AuNPs

The evaporation effect of the AuNPs droplets on the DMF was studied to examine if the evaporation of solvent in the droplet could influence the result of the sensing based on AuNPs. Before this, we first measured the absorbance of the AuNPs droplets with concentrations from 0.25 to 0.05 nM on the DMF. During the experiment, when the droplets moved into the sensing electrode, the light intensity would abruptly increase because of the focusing effect of the droplet (Figure 1a). As the evaporation will increase the concentration of AuNPs in the droplets, the absorbance at approximately 520 nm of the droplets was measured to quantify the evaporation, which is linear to the concentration of the AuNPs (Figure 3a). Since the concentration of a water droplet ($A_H = I_{H20}/I_{base}$) should not change during evaporation, it was used as a reference to eliminate the drift of optical sensor. This method could be useful to investigate the droplet evaporation on the DMF chip, as it provides the dynamic information of the droplets. As shown in Figure 3b, the $\Delta A/A_0$ of the AuNPs droplets with a volume of 5 μL slightly decreased with time, where $\Delta A/A_0 = (A_s/A_H)/(A_{s,0}/A_{H,0}) - 1$ is the change of absorbance as a related error to the baseline at 0 min. The decrease in $\Delta A/A_0$ was less than 0.5% within 10 min, revealing that the evaporation can hardly affect the measurement if the assay lasts several minutes. In a previous study, a complicated process was observed during the evaporation of droplets in an open configuration on a hydrophobic surface [26]. As the height of the droplet changes during evaporation, it has a significant effect on the absorbance. In this study, a two-plate configuration of DMF was used for droplet control. Therefore, the height of the droplet was constant until most of the droplet solvent was dried up, and the concentration of the AuNPs plays the main role in the measurement during evaporation. On the other hand, for detections that endure for longer time, it is feasible to fill the DMF chip with silicon oil to prevent the evaporation of droplets, which can be easily realized in a well-sealed DMF chip [27].

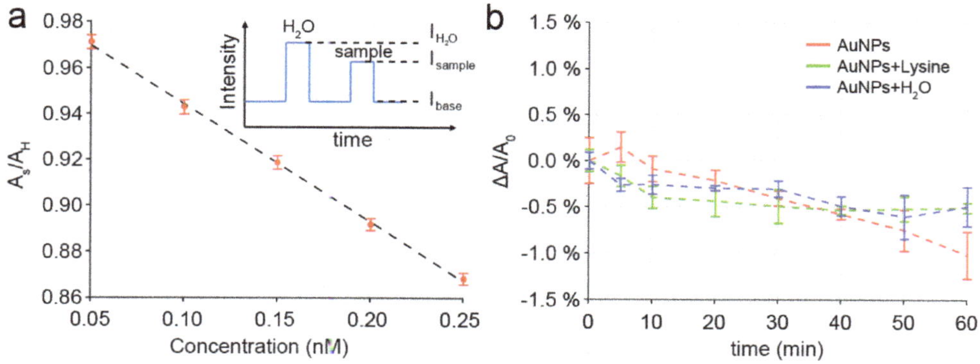

Figure 3. (**a**) The absorbance of droplets is linear to the concentration of the AuNPs (N = 3); Inset: diagram for the evaluation of the related absorbance, $A_H = I_{H2O}/I_{base}$, $A_s = I_{sample}/I_{base}$. (**b**) Changing of the absorbance of AuNPs droplets (5 μL) induced by the evaporation on the DMF chip, where $\Delta A/A_0 = (A_s/A_H)/(A_{s,0}/A_{H,0}) - 1$ is the change of absorbance as a related error to the baseline at 0 min.

3.3. On-Chip Detection of Hg^{2+}

To validate the sensing strategy, a proof-of-principle experiment was conducted by integrating the DMF system with a two-wavelength measuring module. The DMF system was used to automatically detect the concentration of Hg^{2+} based on AuNPs (Figure 1c). The AuNPs mixed with lysine were used as a reagent to detect and quantify the concentration of Hg^{2+}. The principle of the detection is referred to a two-step mechanism by promoting the response of AuNPs to the Hg^{2+} with lysine as previously reported [18], which also has the advantage of high selectivity: First, the Hg^{2+} in the droplet was reduced on the surface of AuNPs to form a shell. Then, the AuNPs covered with Hg^{2+} were aggregated, which was induced by the interaction between lysine and the shell of Hg, resulting in a rapid color change from red to purple due to the effect of the LSPR [19]. During the experiment, sample droplets with different concentrations ranging from 0, 0.05, 0.25, 0.50, 1.00, 2.00, 2.50, 3.00 to 4 μM were loaded on the DMF chip with a volume of 2 μL. The reagent droplet (5 μL of AuNPs buffer and 1 μL of 1 mM lysine) was loaded on the DMF and mixed with the sample droplet. Then, a process for promoting the reaction was conducted by moving the droplets around the chip for 3 min. Mixed droplets were driven to the sensing electrodes for measuring the absorbance at 520 nm and 730 nm (Figure 4a). We define the absorbance ratio as A_{520}/A_{730}, where A_{520} and A_{730} were the related absorbance (A_s/A_H) at 520 nm and 730 nm, respectively. The change of the absorbance ratio of the AuNPs (A_{520}/A_{730}) has a good relation with the concentration of the Hg^{2+} (Figure 4b), which is consistent with the results, as previously reported in bulk solution [18]. A similar trend was observed in the UV/vis absorbance spectra of the mixed droplets with Hg^{2+} concentrations of 0 μM and 2 μM by both measuring a collection of 200 μL mixed droplets. In addition, the limit of detection based on a signal-to-noise ratio of 3:1 was only 0.01 μM, which is lower than the limit of acceptable Hg^{2+} concentration in drinkable water (0.03 μM) according to the WHO. With the increase in the Hg^{2+} concentration, the A_{520}/A_{730} was almost saturated. This is reasonable, since most of the AuNPs would be aggregated at an excessive concentration of Hg^{2+} [20]. The dynamic range can be further optimized by adjusting the concentration of AuNPs, the volume ratio of the AuNPs droplet and the sample droplet, and the concentration of lysine [28].

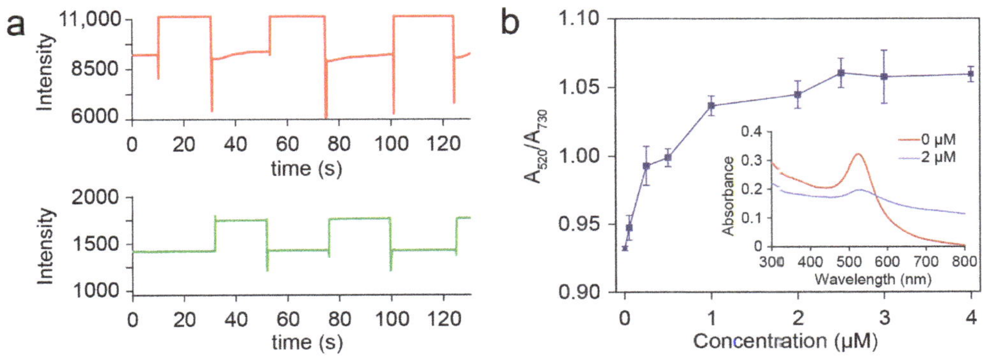

Figure 4. (**a**) Raw data of the absorbance with Hg^{2+} concentration of 2 μM at 730 nm (Red) and 520 nm (Green). (**b**) A_{520}/A_{730} ratio of the AuNPs droplets reacted with Hg^{2+} concentration ranging from 0, 0.05, 0.25, 0.50, 1.00, 2.00, 2.50, 3.00 to 4 μM (inset: the UV/vis absorbance spectra with Hg^{2+} concentration of 0 μM and 2 μM).

4. Conclusions

In summary, we studied the EWOD property of the AuNPs droplets on the DMF chip, which reveals that the interaction of the AuNPs at the solid–liquid interface of a droplet could influence the electrowetting force for the droplet manipulation. The electrowetting contact angle of the AuNPs follows a modified Young–Lippmann equation with an additional item representing the changing of solid–liquid surface tension. Therefore, a higher voltage is needed to manipulate the AuNPs droplets than the water droplets at the same condition. We also found that the evaporation of the solvent in the AuNPs droplets had a limited influence (<0.5%) on the colorimetric sensing within 10 min. As an example experiment, we conducted the AuNPs-based colorimetric sensing of the concentration of Hg^{2+} on the DMF platform. The sensitivity of the detection was modified by introducing a two-wavelength measuring method. The results indicated that the concentration of Hg^{2+} could be determined by using the AuNPs-based method on the DMF, and the limit of detection is as low as 0.01 μM. The proposed method based on AuNPs with the advantages of automation, low reagent cost and small size could be further developed for a wide range of applications such as point-of-care testing of nucleic acid, antibody, proteins, and small molecules.

Author Contributions: Conceptualization, Z.G. and C.K.; methodology, Z.G. and H.-F.W.; validation, Z.G., J.-J.L., L.-W.D. and J.-G.W.; writing—original draft preparation, Z.G. and B.-Y.Y.; writing—review and editing, C.K., J.-G.W. and J.-L.Z. All authors have read and agreed to the published version of the manuscript.

Funding: This research was funded by the National Natural Science Foundation of China (62103148, 31701698), the National Key R&D Program of China (2021YFC2101100) and the financial support from Shanghai Key Laboratory of Forensic Medicine (Academy of Forensic Science, KF1910).

Conflicts of Interest: The authors declare no conflict of interest.

References

1. Zhong, J.; Riordon, J.; Wu, T.C.; Edwards, H.; Wheeler, A.R.; Pardee, K.; Aspuru-Guzik, A.; Sinton, D. When robotics met fluidics. *Lab Chip* **2020**, *20*, 709–716. [CrossRef]
2. Ruan, Q.; Ruan, W.; Lin, X.; Wang, Y.; Zou, F.; Zhou, L.; Zhu, Z.; Yang, C. Digital-WGS: Automated, highly efficient whole-genome sequencing of single cells by digital microfluidics. *Sci. Adv.* **2020**, *6*, eabd6454. [CrossRef]
3. Jin, K.; Hu, C.; Hu, S.; Hu, C.; Li, J.; Ma, H. "One-to-three" droplet generation in digital microfluidics for parallel chemiluminescence immunoassays. *Lab Chip* **2021**, *21*, 2892–2900. [CrossRef]
4. Zhai, J.; Li, H.; Wong, A.H.-H.; Dong, C.; Yi, S.; Jia, Y.; Mak, P.-I.; Deng, C.-X.; Martins, R.P. A digital microfluidic system with 3D microstructures for single-cell culture. *Microsyst. Nanoeng.* **2020**, *6*, 6. [CrossRef] [PubMed]

5. Li, B.B.; Scott, E.Y.; Chamberlain, M.D.; Duong, B.T.V.; Zhang, S.; Done, S.J.; Wheeler, A.R. Cell invasion in digital microfluidic microgel systems. *Sci. Adv.* **2020**, *6*, eaba9589. [CrossRef]
6. Grant, N.; Geiss, B.; Field, S.; Demann, A.; Chen, T.W. Design of a Hand-Held and Battery-Operated Digital Microfluidic Device Using EWOD for Lab-on-a-Chip Applications. *Micromachines* **2021**, *12*, 1065. [CrossRef] [PubMed]
7. Perry, J.M.; Soffer, G.; Jain, R.; Shih, S.C.C. Expanding the limits towards 'one-pot' DNA assembly and transformation on a rapid-prototype microfluidic device. *Lab Chip* **2021**, *21*, 3730–3741. [CrossRef]
8. Alistar, M.; Gaudenz, U. OpenDrop: An Integrated Do-It-Yourself Platform for Personal Use of Biochips. *Bioengineering* **2017**, *4*, 45. [CrossRef] [PubMed]
9. Lee, M.S.; Hsu, W.; Huang, H.Y.; Tseng, H.Y.; Lee, C.T.; Hsu, C.Y.; Shieh, Y.C.; Wang, S.H.; Yao, D.J.; Liu, C.H. Simultaneous detection of two growth factors from human single-embryo culture medium by a bead-based digital microfluidic chip. *Biosens. Bioelectron.* **2020**, *150*, 111851. [CrossRef] [PubMed]
10. Sista, R.S.; Ng, R.; Nuffer, M.; Basmajian, M.; Coyne, J.; Elderbroom, J.; Hull, D.; Kay, K.; Krishnamurthy, M.; Roberts, C.; et al. Digital Microfluidic Platform to Maximize Diagnostic Tests with Low Sample Volumes from Newborns and Pediatric Patients. *Diagnostics* **2020**, *10*, 21. [CrossRef]
11. Li, J.; Ha, N.S.; Liu, T.; van Dam, R.M.; Kim, C.J. Ionic-surfactant-mediated electro-dewetting for digital microfluidics. *Nature* **2019**, *572*, 507–510. [CrossRef]
12. Shamsi, M.H.; Choi, K.; Ng, A.H.C.; Chamberlain, M.D.; Wheeler, A.R. Electrochemiluminescence on digital microfluidics for microRNA analysis. *Biosens. Bioelectron.* **2016**, *77*, 845–852. [CrossRef]
13. Dryden, M.D.; Rackus, D.D.; Shamsi, M.H.; Wheeler, A.R. Integrated digital microfluidic platform for voltammetric analysis. *Anal. Chem.* **2013**, *85*, 8809–8816. [CrossRef] [PubMed]
14. Gu, Z.; Wu, M.L.; Yan, B.Y.; Wang, H.F.; Kong, C. Integrated Digital Microfluidic Platform for Colorimetric Sensing of Nitrite. *ACS Omega* **2020**, *5*, 11196–11201. [CrossRef]
15. Lamanna, J.; Scott, E.Y.; Edwards, H.S.; Chamberlain, M.D.; Dryden, M.D.M.; Peng, J.; Mair, B.; Lee, A.; Chan, C.; Sklavounos, A.A.; et al. Digital microfluidic isolation of single cells for -Omics. *Nat. Commun.* **2020**, *11*, 5632. [CrossRef] [PubMed]
16. Sklavounos, A.A.; Nemr, C.R.; Kelley, S.O.; Wheeler, A.R. Bacterial classification and antibiotic susceptibility testing on an integrated microfluidic platform. *Lab Chip* **2021**, *21*, 4208–4222. [CrossRef]
17. Han, S.; Zhang, Q.; Zhang, X.; Liu, X.; Lu, L.; Wei, J.; Li, Y.; Wang, Y.; Zheng, G. A digital microfluidic diluter-based microalgal motion biosensor for marine pollution monitoring. *Biosens. Bioelectron.* **2019**, *143*, 111597. [CrossRef]
18. Sener, G.; Uzun, L.; Denizli, A. Lysine-Promoted Colorimetric Response of Gold Nanoparticles: A Simple Assay for Ultrasensitive Mercury(II) Detection. *Anal. Chem.* **2013**, *86*, 514–520. [CrossRef] [PubMed]
19. Gu, Z.; Jing, C.; Ying, Y.L.; He, P.; Long, Y.T. In situ high throughput scattering light analysis of single plasmonic nanoparticles in living cells. *Theranostics* **2015**, *5*, 188–195. [CrossRef]
20. Cao, Y.; Wu, J.; Pang, B.; Zhang, H.; Le, X.C. CRISPR/Cas12a-mediated gold nanoparticle aggregation for colorimetric detection of SARS-CoV-2. *Chem. Commun.* **2021**, *57*, 6871–6874. [CrossRef] [PubMed]
21. Pramanik, A.; Gao, Y.; Patibandla, S.; Mitra, D.; McCandless, M.G.; Fassero, L.A.; Gates, K.; Tandon, R.; Chandra Ray, P. The rapid diagnosis and effective inhibition of coronavirus using spike antibody attached gold nanoparticles. *Nanoscale Adv.* **2021**, *3*, 1588–1596. [CrossRef]
22. Orejon, D.; Sefiane, K.; Shanahan, M.E.R. Young-Lippmann equation revisited for nano-suspensions. *Appl. Phys. Lett.* **2013**, *102*, 201601. [CrossRef]
23. Li, L.; Gu, Z.; Zhou, J.-L.; Yan, B.; Kong, C.; Wang, H.; Wang, H.-F. Intelligent droplet tracking with correlation filters for digital microfluidics. *Chin. Chem. Lett.* **2021**, *11*. [CrossRef]
24. Haiss, W.; Thanh, N.T.K.; Aveyard, J.; Fernig, D.G. Determination of Size and Concentration of Gold Nanoparticles from UV−Vis Spectra. *Anal. Chem.* **2007**, *79*, 4215–4221. [CrossRef] [PubMed]
25. Kumar, S.; Kumar, P.; DasGupta, S.; Chakraborty, S. Electrowetting of a nano-suspension on a soft solid. *Appl. Phys. Lett.* **2019**, *114*, 073702. [CrossRef]
26. Chandramohan, A.; Chakraborty, M.; Weibel, J.A.; Garimella, S.V. Evaporation-Driven Micromixing in Sessile Droplets for Miniaturized Absorbance-Based Colorimetry. *ACS Omega* **2019**, *4*, 22385–22391. [CrossRef]
27. Qiu, W.; Nagl, S. Automated Miniaturized Digital Microfluidic Antimicrobial Susceptibility Test Using a Chip-Integrated Optical Oxygen Sensor. *ACS Sens.* **2021**, *6*, 1147–1156. [CrossRef]
28. Zhang, X.; Fan, X.; Wang, Y.; Lei, F.; Li, L.; Liu, J.; Wu, P. Highly Stable Colorimetric Sensing by Assembly of Gold Nanoparticles with SYBR Green I: From Charge Screening to Charge Neutralization. *Anal. Chem.* **2019**, *92*, 1455–1462. [CrossRef]

micromachines

MDPI

Article

Microfluidic Chip with Two-Stage Isothermal Amplification Method for Highly Sensitive Parallel Detection of SARS-CoV-2 and Measles Virus

Qin Huang [1,†], Xiaohui Shan [1,†], Ranran Cao [2,†], Xiangyu Jin [1], Xue Lin [1], Qiurong He [3], Yulei Zhu [3], Rongxin Fu [1], Wenli Du [1], Wenqi Lv [1], Ying Xia [3,*] and Guoliang Huang [1,*]

1 Department of Biomedical Engineering, School of Medicine, Tsinghua University, Beijing 100084, China; xiaoqingzi12345@163.com (Q.H.); 17763384805@163.com (X.S.); 18810913852@163.com (X.J.); ruixuezhaofn@yeah.net (X.L.); thu_frx@tsinghua.edu.cn (R.F.); wenlidu1230@163.com (W.D.); 18254369766@163.com (W.L.)

2 Sichuan Center for Disease Control and Prevention, Chengdu 610041, China; ranran.cao@foxmail.com

3 West China School of Public Health and West China Fourth Hospital, Sichuan University, Chengdu 610041, China; heqiurong2009@163.com (Q.H.); yulei_zhu@yeah.net (Y.Z.)

* Correspondence: xiayingscu@scu.edu.cn (Y.X.); tshgl@tsinghua.edu.cn (G.H.); Tel.: +86-(010)-62797213 (G.H.)

† Contributed equally as co-first authors.

Abstract: A two-stage isothermal amplification method, which consists of a first-stage basic recombinase polymerase amplification (RPA) and a second-stage fluorescence loop-mediated isothermal amplification (LAMP), as well as a microfluidic-chip-based portable system, were developed in this study; these enabled parallel detection of multiplex targets in real time in around one hour, with high sensitivity and specificity, without cross-contamination. The consumption of the sample and the reagent was 2.1 μL and 10.6 μL per reaction for RPA and LAMP, respectively. The lowest detection limit (LOD) was about 10 copies. The clinical amplification of about 40 nasopharyngeal swab samples, containing 17 SARS-CoV-2 (severe acute respiratory syndrome coronavirus) and 23 measles viruses (MV), were parallel tested by using the microfluidic chip. Both clinical specificity and sensitivity were 100% for MV, and the clinical specificity and sensitivity were 94.12% and 95.83% for SARS-CoV-2, respectively. This two-stage isothermal amplification method based on the microfluidic chip format offers a convenient, clinically parallel molecular diagnostic method, which can identify different nucleic acid samples simultaneously and in a timely manner, and with a low cost of the reaction reagent. It is especially suitable for resource-limited areas and point-of-care testing (POCT).

Keywords: two-stage isothermal amplification; microfluidic chip; parallel molecular diagnostic; SARS-CoV-2; measles virus

Citation: Huang, Q.; Shan, X.; Cao, R.; Jin, X.; Lin, X.; He, Q.; Zhu, Y.; Fu, R.; Du, W.; Lv, W.; et al. Microfluidic Chip with Two-Stage Isothermal Amplification Method for Highly Sensitive Parallel Detection of SARS-CoV-2 and Measles Virus. *Micromachines* **2021**, *12*, 1582. https://doi.org/10.3390/mi12121582

Academic Editors: Stefano Zampolli and Aiqun Liu

Received: 19 October 2021
Accepted: 16 December 2021
Published: 19 December 2021

1. Introduction

The recognition of infection pathogens is crucial for the assessment of disease state and treatment efficacy; therefore, diagnosing infectious diseases accurately and in a timely manner is important, especially for diseases caused by multiple pathogens [1–3]. For example, pneumonia can be caused by a variety of organisms, including bacteria, bacteria-like organisms, viruses and fungi [4,5]. Antibiotics are usually used to treat bacterial pneumonia, but inappropriate prescribing of antibiotics (for example, for viral infections) may result in drug resistance [6]. However, it takes time to identify the type of pathogen causing pneumonia and to choose the best antibiotic to treat it. The most common method to confirm an infection is a blood test [7], which can identify the type of organism. However, it is impossible for precise identification, such as telling which kind of virus or bacteria caused the infection. Culture-based microbiological tests, which are verified as the golden standard method for accurate pathogen identification, have suffered from a delay between sampling and result reporting, ranging from hours to days; furthermore, this method also requires a skilled experimenter and a facilitated laboratory [8].

Although molecular diagnostic methods, especially PCR, have developed rapidly throughout the last several decades [9], the clinical application of these methods was still relatively limited until the year 2020. Since the emergence of a respiratory disease caused by a new kind of coronavirus, SARS-CoV-2, there have been more than 229 million confirmed cases of COVID-19 [10]. Due to the fact that molecular diagnostic methods can detect a large number of pathogens within a greatly shortened time (usually 2 to 3 h) with accuracy, molecular diagnostics have become the gold standard methods for SARS-CoV-2 testing, particularly the nucleic acid test method [11]. However, the test is only available in fully equipped first-level hospitals. The application in community clinics and clinics in villages and towns, especially in resource-limited areas and countries, is still hampered because the thermal cycling (PCR) process complicates the instrumentation; this means a high cost of equipment and a need for professional molecular experimenters. Moreover, most of the existing nucleic acid molecular diagnostic assays can only detect one single target in a test, which is limited in scope. Although multiplex PCR (mPCR) has the ability to amplify many nucleic acid (NA) targets in a single reaction [12,13], it requires probes with different color fluorophores for multiple detection, which complicates the assay and increases the cost [14].

In recent years, isothermal amplification methods with no need for a temperature controlling component, such as loop-mediated isothermal amplification (LAMP) [15], recombinase polymerase amplification (RPA) [16], and nucleic acid sequence-based amplification (NASBA) [17], have attracted interest among researchers [18–21]. A few groups have developed multiplexed RPA [22,23] and multiplexed LAMP assays [24–27] for co-detection of a small number (i.e., $\leq$4) of targets. There are also some isothermal amplification methods based on microfluidic chips that have been developed for rapid detection and identification of pathogens [28–31].

Specifically, research based on isothermal amplification methods have been reported for the detection of SARS-CoV-2. As reported by Kim Y. et al. (Preprint), they reported a single-strand RPA for visual detection of SARS-CoV-2 [32], while Zhang Y. developed a AuNP-based colorimetric visual assay, which is called Cas12a-assisted RT-LAMP/AuNP (CLAP), for on-site diagnosis of COVID-19 [33]. El-Tholoth M. also used a two-stage isothermal amplification (Penn-RAMP), which was developed by Song J. in 2017 [34] for the detection of SARS-CoV-2 [35]. However, these methods were performed in tubes, which means large volume (25 to 50 μL) and the detection of only one target in a single tube.

To enhance the limits of detection, improve specificity, and enable parallel detection of multiple nucleic acid targets, we proposed a two-stage isothermal amplification, with an integrated microfluidic chip for the detection of different types of pathogens in a closed system, with a volume of 1 μL RPA, plus 10 μL LAMP for each target, during about one hour process; this consists of first-stage basic RPA and second-stage real-time LAMP, while products of RPA serve as templates for LAMP. The performance of the two-stage isothermal method was evaluated by comparing the limit of detection and reproducibility with basic RPA and fluorescence LAMP. The microfluidic chip's capability was demonstrated by comparing the sensitivity and reproducibility of two-stage isothermal reactions on the microfluidic chip with in tubes. The ability to perform parallel detection of different types of targets on one chip was examined by detecting 40 clinical samples and comparing the results with traditional microbiological methods. The results led us to believe that the closed microfluidic chip system with the two-stage isothermal amplification method has high sensitivity, specificity and reproducibility, and consumes small volume reaction components with a fast reaction speed, which is suitable for real-time multiple pathogen diagnosis in a single detection run.

2. Materials and Methods

2.1. Targets and Sample Preparation

Two viruses were targeted in this study: (1) SARS-CoV-2 (severe acute respiratory syndrome coronavirus), which is a pathogen that caused the outbreak of coronavirus disease 2019 (COVID-19) worldwide, and (2) Measles Virus (MV), which is an enveloped negative-strand RNA virus that causes measles and is highly contagious. Infected patients of both of these pathogens develop high fever and cold-like symptoms at an early stage, and can experience pneumonia or even death if the patient is elderly, very young, or immunocompromised.

To detect and distinguish these two viruses from different infectious diseases, we used two different types of samples: (1) Plasmid DNA containing the targeted sequences, which were synthesized at Sangon Biotech (Shanghai, China) to mimic the SARS-CoV-2 and MV target, (2) clinical nasopharyngeal swab samples from SARS-CoV-2 and MV infected patients. Purified MV RNA and 40 nasopharyngeal swab samples from patients infected by COVID-19 or MV were obtained from the Center for Disease Control and Prevention (Sichuan, China). All clinical samples were heated at 98 °C for 15 min before nucleic acid extraction, according to biosafety laboratory requirements. RNA nucleic acid from Clinical samples was extracted using the Maxwell 16 Viral Total Nucleic Acid Purification Kit (Promega Co., Madison, WI, USA) at a Biological safety protection third-level laboratory (P3). The prepared nucleic acids were immediately stored at −20 °C until use.

2.2. RPA and LAMP Reaction in Tube

We performed the recombinase polymerase amplification (RPA) reaction for the first-stage amplification with the TwistAmp® Basic Kit and TwistAmp® Exo Kit (TwistDx Inc., Cambridge, UK). Targeting RNA experiments were conducted by adding 0.5 U/µL Avian Myeloblastosis Virus Reverse Transcriptase (AMV RT) (Promega Co., Madison, WI, USA) to synthesize a cDNA at first. The second-stage amplification of the loop-mediated isothermal amplification (LAMP) reaction was performed with a WarmStart LAMP Kit (New England BioLabs Inc., Beijing, China). The product of the RPA reaction was purified with a Universal DNA Purification Kit (Tiangen, Beijing, China).

The sequences for the primers are listed in Table 1. The RPA assays required a set of two primers for each target. The RPA primers were designed according to TwistAmp® DNA amplification kits Combined instruction manual. LAMP assays required a set of six primers to amplify a specific target, including two outer primers (F3 and B3), two inner primers (forward inner primer (FIP) and backward inner primer (BIP)), and two loop primers (forward loop primer (LF) and backward loop primer (LB)). All LAMP primers were designed on the official primer design website (https://primerexplorer.jp/elamp4.0.0/index.html), and their specificities were verified by Basic Local Alignment Search Tool (BLAST) analysis (Version 4, Bethesda, MD, USA). All primers (as listed in Table 1) were ordered through Sangon Biotech (Shanghai, China).

The RPA and LAMP reactions in tubes were performed by using a real-time thermoscycler (CFX96, Bio-Rad, Hercules, CA, USA). The first-stage RPA reaction was set at 39 °C for 15 min, then 1 µL aliquots of the RPA product were used as templates for the second-stage LAMP reaction, which was set at 65 °C for 50 min. The amplification of LAMP was monitored in real time by detecting its fluorescence signal.

2.3. Air-Insulated Microfluidic Chip Design

Our air-insulated, disc-type microfluidic chip had a diameter of 62 mm and a thickness of 3.0 mm. The chip was designed with AutoCAD 2017. The chip was made of polymethyl methacrylate (PMMA) and fabricated by computer-controlled (CNC) milling. The structure of the chip is illustrated in Figure 1. It consisted of four layers: a basement layer carved with all the microstructures, two cover layers for encapsulation, and a film to seal the inlet holes. The basement layer consisted of two sides: side A was carved with microstructures for the RPA reaction, while side B was carved for the LAMP reaction. Side A consisted of a V-shape quantitative channel (width: 0.6 mm, depth: 0.4 mm), an air balance channel (width: 0.1 mm, depth: 0.1 mm), inlet A and inlet B. Side B consisted of a siphon valve (width: 0.3 mm, depth: 0.3 mm), a storage chamber (depth: 1.5 mm), a quantitative chamber (depth: 1.3 mm) and amplification chambers (depth: 1.5 mm). The V-shape channel was used to divide the RPA mix into quantitative chambers, with 2.10 μL of liquid distributed into each chamber. The quantitative chamber, with a volume of about 8.70 μL, was used for the RPA reaction and the quantification of the LAMP mix. The amplification chambers for the LAMP reactions had a diameter of 3.0 mm and a volume of 10.60 μL. The number of amplification chambers could be adjusted according to the number of targets and controls. The chip, illustrated in Figure 1, contained two identical parts, which could perform two samples on one chip for one detection. One hard polycarbonate (PC) film was used to cover side A, and one soft PC film was used to cover side B. Both PC films were tightly adhered to both sides of the chip to form an enclosed microfluidic chip.

Table 1. Primers for SARS-CoV-2 and MV.

Target	Primer Name	Primer Sequence (5′-3′)
SARS-CoV-2 (CoV-2)	CoV-RPA-F	ATACACTAATTCTTTCACACGTGGTGTTTA
	CoV-RPA-R	AGTAGCGTTATTAACAATAAGTAGGGACTGGG
	CoV-LAMP-F3	ACACTAATTCTTTCACACGTGGTG
	CoV-LAMP-B3	ATTAACAATAAGTAGGGACTGGG
	CoV-LAMP-FIP	CCAGAGACATGTATAGCATGGAACCCATTCAACTCAGGACTTGTTCT
	CoV-LAMP-BIP	GAGGTTTGATAACCCTGTCCTACCATCTTCGAATCTAAAGTAGTACC
	CoV-LAMP-LF	CATTGGAAAAGAAAGGTA
	CoV-LAMP-LB	TGCTTCCACTGAGAAG
Measles Virus (MV)	MV-RPA-F	AGAATAATGAAGAAGGGGGAGACTATTATGA
	MV-RPA-R	CAGCAGCAGCTGTCCTCTGGAACTGGTCCG
	MV-LAMP-F3	GGACACCTCTCAAGCATC
	MV-LAMP-B3	CAGCAGCTGTCCTCTGGAA
	MV-LAMP-FIP	CGGCCTGAATCTCTGCCTATGATTGGGAAGGATCCCAACG
	MV-LAMP-BIP	GTTCTCAAGAAACCCGCTGCCCTGGTCCGTCCATTTGTCA
	MV-LAMP-LF	GGATTGAGTTCGACATCTGC
	MV-LAMP-LB	AGCCGACAACTCCAAGGA

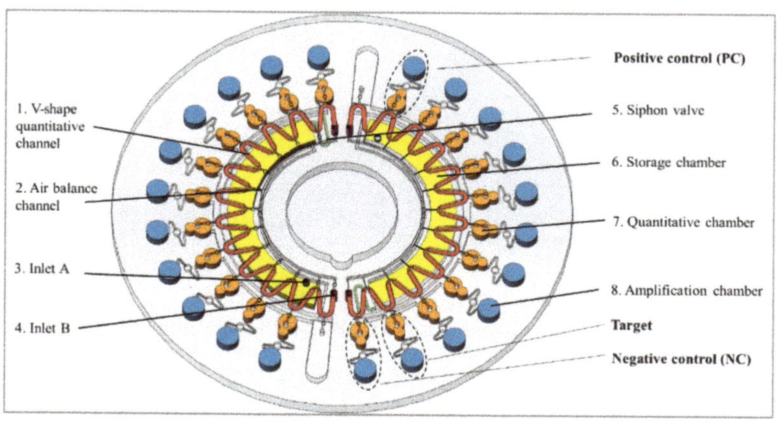

Figure 1. Three dimensional structure of the microfluidic chip.

2.4. RPA and LAMP Reactions on Microfluidic Chip

On the microfluidic chip, the reactions were performed with the same kits as in the tube, and with the same proportions of the components. After the basement layer was cleaned with 70% ethanol, the quantitative chamber was preloaded with RPA primers and $Mg(OAc)_2$, and the amplification chamber was embedded with LAMP primers designed to identify one pathogen using low melting-point Sepharose CL-4B. The siphon valve was modified to be hydrophilic using the hydrophilic reagent (1% agarose in water). The chip was sealed with two PC films and compressed by a press machine to form an enclosed microfluidic chip. After that, the chip was stored at 4 °C until use.

For chip use, the RPA mix (25 μL) and LAMP mix (120 μL) were added separately into the V-shape channel and storage chamber using a pipette, through inlets A and B, respectively. After that, all of the inlets and outlets were sealed with adhesive tape. Figure 2 illustrates the entire flow control of the chip. Next we put the chip into the portable analyzer (Authorized patent ZL201110113608.1) which can conduct fully automate nucleic acid analysis. The entire flow control of the chip, on-chip RPA and LAMP reaction, and real-time fluorescent signal of LAMP amplification detection were all completed in the analyzer. The on-chip RPA was performed at 39 °C for 10 min, and on-chip LAMP was performed at 65 °C for 50 min.

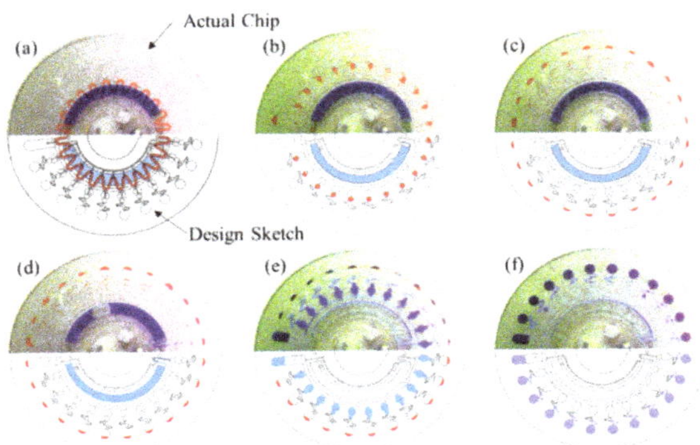

Figure 2. Illustration of the entire flow control of the chip. (**a**) The initial state of the chip with the RPA mix (red dye) and LAMP mix (blue dye), dried primers; (**b**) RPA mix was divided into quantitative chambers at 2000 rpm for 30 s; (**c**) the RPA product was transferred into amplification chamber as template for LAMP; (**d**) LAMP mix was primed into the siphon valve by the capillary action at 100 rpm for 30 s; (**e**) LAMP mix was transferred into the separated sub-volumes (10.6 μL per chamber) at 2000 rpm for 30 s; (**f**) LAMP mix was distributed into reaction chambers at 6000 rpm for 60 s.

3. Results

3.1. Comparison of Sensitivity and Reproducibility of Basic RPA Followed by Fluorescence LAMP (bRPA-LAMP) with Basic RPA and Fluorescence LAMP in Tube

To demonstrate the sensitivity of the bRPA-LAMP assay, we compared it with basic RPA and fluorescence LAMP in the tube. The targets were 1E+4, 1E+3, 1E+2, and 1E+1 copies of MV RNA. Three duplicate reactions were performed for each nucleic acid template concentration to evaluate reproducibility. The sensitivity of basic RPA was measured with agarose gel (1.5%) electrophoretograms of purified RPA products that were cleaned up with a DNA product purification kit (Tiangen, Beijing, China). The sensitivity of fluorescence LAMP and bRPA-LAMP were measured with the time of positive value (Tp), which was defined as the time at the second derivative inflexions of the exponential DNA amplification curves.

As Figure 3 indicates, basic RPA measured by agarose gel (1.5%), electrophoretograms had the lowest sensitivity with a detection limit of 1E+4 copies of MV RNA. The LAMP process alone produced a detectable signal with 1E+3 copies, but had 2 positive signals out of three repeats for 1E+2 copies of MV RNA. In contrast, the bRPA-LAMP method yielded a significantly higher sensitivity than LAMP alone. The unpurified RPA product served as a template for fluorescence LAMP successfully producing a signal with samples containing 1E+2 copies of MV RNA. It is noteworthy that the purified RPA product that served as a template for fluorescence LAMP can detect samples containing 1E+1 copies of MV RNA as early as about 21.3 min. Its positive signal was produced even earlier than the LAMP with 1E+3 copies of MV RNA template. As Table 2 shows, the Tp values of RPA followed by fluorescence LAMP were significantly smaller than those of LAMP alone. The corresponding standard deviation (SD) values indicate that RPA followed by fluorescence LAMP, as well as LAMP, was reasonably reproducible (~4% relative standard deviation in the Tp values). Thus, we can see that the purified RPA product followed by LAMP had the highest sensitivity of 1E+1 copies, followed by the unpurified RPA product, followed by LAMP with a sensitivity of 1E+2 copies. Furthermore, LAMP had a sensitivity of 1E+3 copies, while basic RPA had the lowest sensitivity of 1E+4 copies.

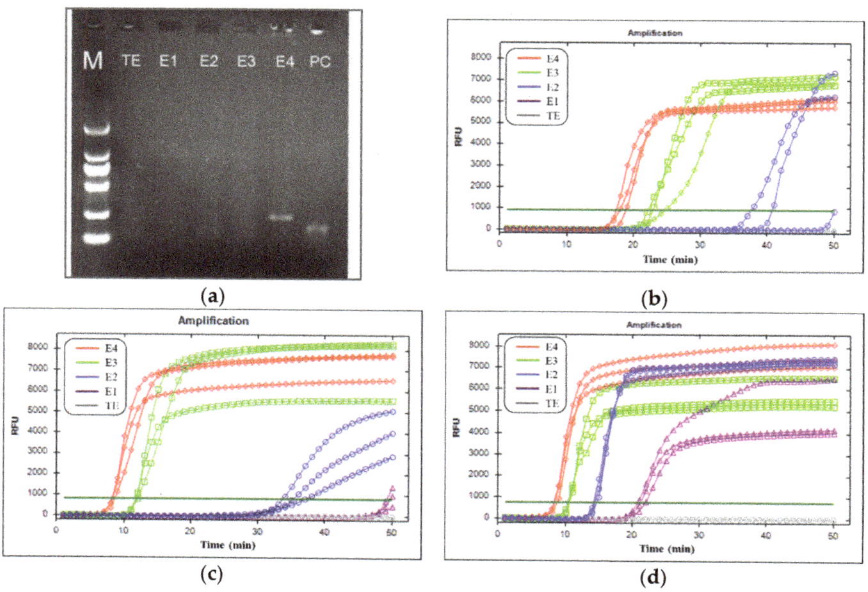

Figure 3. Comparison of sensitivity of bRPA-LAMP with basic RPA and fluorescence LAMP for MV RNA. (**a**) Agarose gel (1.5%) electrophoretogram image of purified basic RPA products; (**b**) real-time curves of LAMP; (**c**) real-time curves of unpurified RPA product followed by LAMP; (**d**) real-time curves of purified RPA product followed by LAMP. Note: PC: positive control.

Table 2. Tp values and SD of bRPA-LAMP and LAMP alone with 1E+4, 1E+3, 1E+2, and 1E+1 copies of MV RNA as target (N = 3). NS stands for non-detectable signal during the duration of the experiment.

	Tp ± SD (min, N = 3)			
	1E+4 Copies	**1E+3 Copies**	**1E+2 Copies**	**1E+1 Copies**
LAMP	17.8 ± 0.6	22.8 ± 0.9	2 in 3 positive	NS
Unpurified RPA product followed by LAMP	8.4 ± 0.2	11.8 ± 0.2	35.3 ± 1.4	2 in 3 positive
Purified RPA product followed by LAMP	8.3 ± 0.1	10.4 ± 0.1	14.4 ± 0.2	21.3 ± 0.5

3.2. Comparison of Sensitivity of bRPA-LAMP with LAMP on the Microfluidic Chip

We examined the performance of bRPA-LAMP on our microfluidic chip by comparing it with LAMP. MV was targeted by using its RNA at concentrations of 1E+3, 1E+2, and 1E+1 copies. There were 3 duplicate target bioreactors for bRPA-LAMP on one half of the microfluidic chip, and 3 duplicate target bioreactors for LAMP on the other half of the same chip. Reproducibility was evaluated by measuring the standard deviation of the Tp values of 3 repeats, as shown in Table 3.

Table 3. Tp values and SD of bRPA-LAMP and LAMP with concentration of 1E+3, 1E+2, and 1E+1 copies of MV RNA on microfluidic chip (N = 3).

	Tp ± SD (min, N = 3)		
	1E+3 Copies	**1E+2 Copies**	**1E+1 Copies**
bRPA-LAMP	12.33 ± 0.08	17.26 ± 0.07	21.02 ± 0.41
LAMP	15.07 ± 0.02	21.77 ± 0.46	1 in 3 positive

Figure 4 depicts amplification curves with various concentrations of MV RNA. For the nucleic acid concentration at 1E+1 copies, bRPA-LAMP successfully detected the target at about 21 min, while LAMP only generated detectable signal for one chamber out of three duplicates. When the MV RNA concentration is 1E+2 copies, the Tp values are about 17.3 min and 21.8 min for bRPA-LAMP and LAMP, respectively. As for MV RNA at a concentration of 1E+3 copies, bRPA-LAMP produced signals about 3 min earlier than LAMP. For reproducibility, there was no big difference between bRPA-LAMP and LAMP, because they both had a SD of less than 0.5 min.

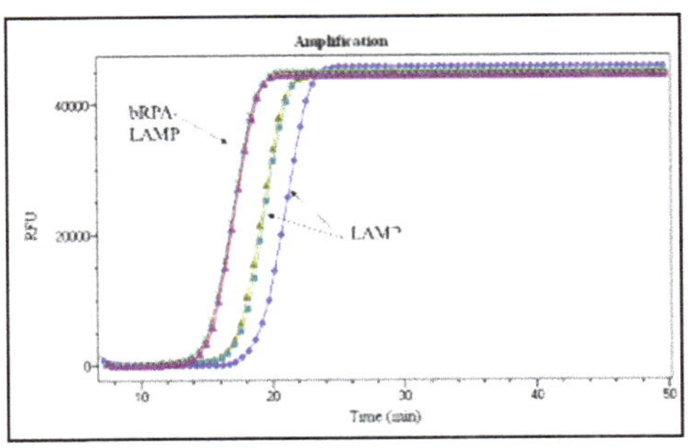

(a)

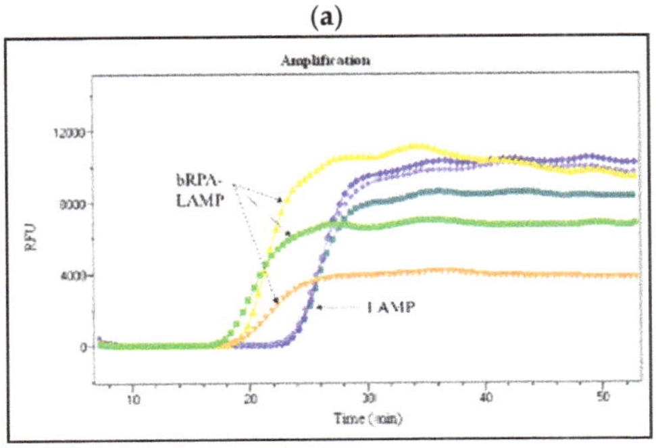

(b)

Figure 4. *Cont.*

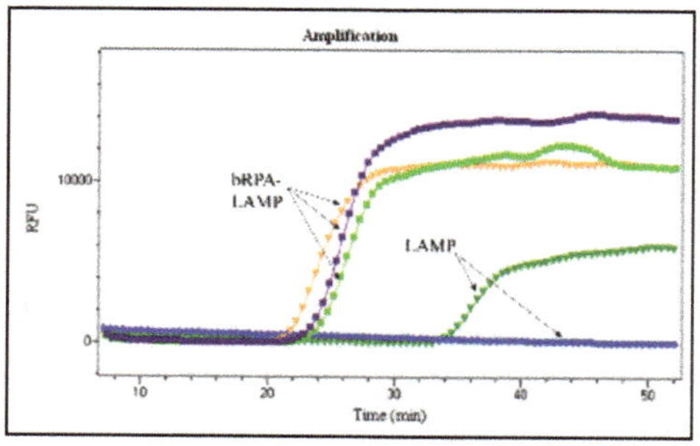

(c)

Figure 4. Comparison of sensitivity of bRPA-LAMP with LAMP on the microfluidic chip. (**a**) Real-time amplification curves for MV RNA at concentrations of 1E+3 copies; (**b**) real-time amplification curves for MV RNA at concentrations of 1E+2 copies; (**c**) real-time amplification curves for MV RNA at concentrations of 1E+1 copies.

When comparing the performance of bRPA-LAMP and LAMP on the microfluidic chip with the same reactions in the EP tubes, regardless of the fact that the reactions on the chip had a longer detection time for template concentration at 1E+3 copies, both methods had better performance for low template concentration (1E+1 and 1E+2 copies); this indicates higher sensitivity for low nucleic acid concentration samples on the microfluidic chip than in the tube.

3.3. Clinical Sensitivity and Specificity of Parallel Detection for Virus on the Same Microfluidic Chip

A total of 40 nasopharyngeal swab samples were classified by standard culture and PCR method, and were detected by using the parallel detection microfluidic chip. The 40 clinical samples included 23 MV positive samples, and 17 SARS-CoV-2 positive samples. Samples that contained one positive target were also used as negative samples for other targets.

To test the parallel detection ability of the microfluidic chip, due to the possibility of multiple infection for respiratory diseases, we mixed nucleic acid extracted from these two kinds of clinical nasopharyngeal swab samples. Figure 5 indicates the result of parallel detection of SARS-CoV-2 and MV at the same time for the mixed nucleic acid sample. The Tp value for clinical samples is longer than that of plasmid DNA, because RNA needs some time to translate its RNA to cDNA.

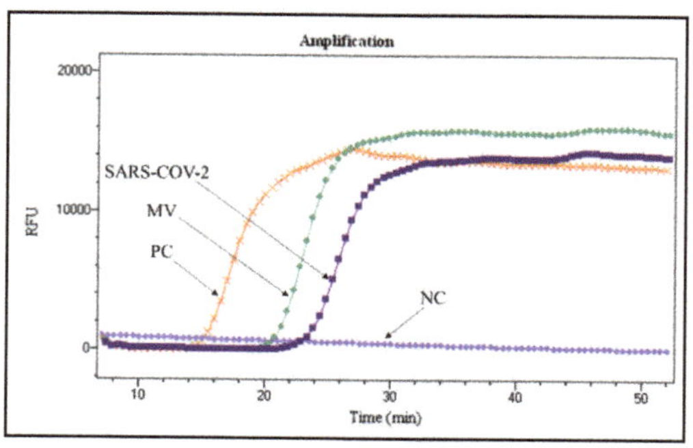

Figure 5. Real-time parallel detection for nasopharyngeal swab samples of SARS-CoV-2 and MV by using bRPA-LAMP on the same microfluidic chip. Note: PC: positive control; NC: negative control.

To evaluate the clinical sensitivity and specificity of different targets, we also tested these samples separately. Table 4 shows that there was no cross-contamination between different targets on the microfluidic chip, with clinical specificity of 95.83% and 100% for SARS-CoV-2 and MV, respectively; this indicates the ability to perform parallel detection of multiple targets, by using one microfluidic chip and one test. Clinical sensitivity was 94.12% and 100% for SARS-CoV-2 and MV, respectively. We also obtained similar clinical sensitivity for SARS-CoV-2 by using the PCR method. The one SARS-CoV-2 positive sample that we failed to detect had a very low pathogenic concentration (<1E+1 copies). The fact that the microfluidic chip did not show a 100% clinical sensitivity for SARS-CoV-2 pointed the way of further improvement for our work, such as RPA product purification and nucleic Acid concentration.

Table 4. Test of the specificities of the microfluidic chip to SARS-CoV-2 and MV RNA in nasopharyngeal swab samples.

Targets		Culture	On Microfluidic Chip	Clinical Sensitivity	Clinical Specificity
SARS-CoV-2	Positive	17	16	94.12%	95.83%
	Negative	23	24		
MV	Positive	23	23	100.0%	100.0%
	Negative	17	17		

4. Discussion

A series of work about microfluidic chips and parallel detection systems has been conducted by our group, including microfluidic chips made with different types of materials; with several designs of micro channels; and targeting bacteria that causes diseases and four types of Ebola virus species, by using separate microfluidic chips [36–38]. We also developed a portable nucleic acid analyzer (used in this study), which is even more suitable than ABI 7500 for trace samples, for the processing and detection of our corresponding microfluidic chips. The microfluidic chip-based isothermal detection system in this study was an integration and optimization of our previous research on microfluidic chips, while extending its clinical amplification extensively.

(1) Detecting different types of nucleic acid targets on one chip.

We made an effort to combine different types of nucleic acid (DNA and RNA) detection in one pot. This is convenient for the detection of clinical samples that may include bacteria and RNA viruses at the same time, such as nasopharyngeal swab samples. Correct classification is important for the diagnosis and treatment of patients. We formerly had to detect these samples twice by using one assay for bacteria and another for viruses, but now we can obtain the results through one test. This is good news for both patients and doctors. Doctors can obtain the infection statement as soon as possible and prescribe the expected treatment, and patients can save money and obtain the right treatment in a timely manner. Due to the fact that it will take some time to transform the RNA nucleic acid into cDNA for later amplification and signal detection, the limit of detection and processing time will be slightly longer than to detect DNA alone; however, the sensitivity of our system for MV RNA can still reach 1E+2 copies, and we are conducting further research to achieve a sensitivity of 1E+1 copies of RNA templates.

(2) Fast detection with high sensitivity and specificity.

This is our first attempt to combine two kinds of isothermal amplification methods, which merges the advantages of RPA and LAMP, while circumventing their shortcomings. RPA can produce 109 times the products in 30 min, but it has a tendency to produce spurious amplicons from primer-dimer or primer-probe complexes [39]. In fact, we also tried fluorescence RPA (probes see Table 1) in this study; however, we obtained similar false positive results for negative controls, as primers will produce dimers at low temperature.

LAMP is highly specific, but its sensitivity is slightly inferior to RPA. By processing a first-stage RPA to reproduce a large number of targets in a short period of time, and a second-stage fluoresce LAMP to detect targets specifically, we can reach a sensitivity of about 10- and 100-fold better than LAMP alone, when operating with unpurified samples and purified RPA products, respectively.

(3) Accurate parallel detection of multiple targets with a small volume.

The microfluidic chip format lays a foundation for the capability of parallel detection (as many as 22 targets) for general pathogens (including DNA and RNA) in around one hour. The risk of contamination for multiple target detection has been minimized for two reasons: (1) products of first-stage RPA can be centrifuged, and diffuse into the second-stage fluorescence LAMP reaction components by mechanical force in a closed system, and (2) all of the LAMP primers have been planted in bioreactors in advance, so there is no need to expose the first-stage RPA products, which are rich in nucleic acid, and LAMP primers, which are rich in primers, to the environment. Moreover, it is cost effective to employ the chip method rather than the traditional tube method. While it consumes 50 μL and 25 μL reagents for RPA and LAMP, respectively, in the tube for each target, the reaction volume on the chip (2.1 μL for RPA and 10.6 μL for LAMP) is approximately 23.8 times and 2.4 times smaller than that in the tube, and thus the consumption of reaction components is largely reduced. In addition, the two-stage isothermal amplification microfluidic chip assay only requires a portable real-time fluorescence detector for reaction and result reporting, which is much more suitable than large-scale PCR amplification instruments in terms of simplicity and convenience.

In summary, the two-stage isothermal amplification method—which consists of a first-stage basic RPA and a second-stage fluorescence LAMP, while implemented in a microfluidic chip format—is capable of detecting multiplex targets (including DNA and RNA) in a short time, with high sensitivity and specificity. The technology and methods in this paper are of value to develop a fully integrated diagnosis system, especially for implementation in POCT.

Author Contributions: Conceptualization, Q.H. (Qin Huang) and X.S.; methodology, Q.H. (Qin Huang), X.S. and R.C.; validation, X.J. and Q.H. (Qiurong He); formal analysis, Q.H. (Qin Huang); investigation, Q.H. (Qiurong He); resources, R.C.; data curation, Q.H. (Qin Huang) and X.S.; writing—original draft preparation, Q.H. (Qin Huang), X.S. and R.C.; writing—review and editing, Y.X., Y.Z., X.L., R.F., W.D. and W.L.; visualization, Q.H. (Qin Huang) and G.H.; supervision, G.H. and Y.X.; project administration, Q.H. (Qin Huang); funding acquisition, G.H. All authors have read and agreed to the published version of the manuscript.

Funding: This research was funded by Tsinghua University Spring Breeze Fund (20201080510), Sichuan Science and Technology Program (2021YFQ0060), National Natural Science Foundation of China (61927819, 81827808, 20170012, 62105177), National Key Research and Development Program of China (2018YFA0704004), Beijing Lab Foundation (20201550018), and Tsinghua Autonomous Research Foundation (20194180031, 20201080058).

Conflicts of Interest: The authors declare no conflict of interest.

References

1. Confield, L.R.; Black, G.P.; Wilson, B.C.; Lowe, D.J.; Theakstone, A.G.; Baker, M.J. Vibrational spectroscopic analysis of blood for diagnosis of infections and sepsis: A review of requirements for a rapid diagnostic test. *Anal. Methods* **2021**, *13*, 157–168. [CrossRef] [PubMed]
2. Lamy, B.; Sundqvist, M.; Idelevich, E.A. ESCMID Study Group for Bloodstream Infections, Endocarditis and Sepsis (ESGBIES). Bloodstream infections—Standard and progress in pathogen diagnostics. *Clin. Microbiol. Infect.* **2020**, *26*, 142–150. [CrossRef] [PubMed]
3. Balsalobre-Arenas, L.; Alarcón-Cavero, T. Rapid diagnosis of gastrointestinal tract infections due to parasites, viruses, and bacteria. *Enferm. Infecc. Microbiol. Clin.* **2017**, *35*, 367–376. [CrossRef] [PubMed]
4. Ning, G.; Wang, X.; Wu, D.; Yin, Z.; Li, Y.; Wang, H.; Yang, W. The etiology of community-acquired pneumonia among children under 5 years of age in mainland China, 2001–2015: A systematic review. *Hum. Vaccines Immunother.* **2017**, *13*, 2742–2750. [CrossRef] [PubMed]

5. Galván, J.M.; Rajas, O.; Aspa, J. Review of Non-Bacterial Infections in Respiratory Medicine: Viral Pneumonia. *Arch. Bronconeumol.* **2015**, *51*, 590–597. [CrossRef] [PubMed]
6. Levy, S.B.; Marshall, B. Antibacterial resistance worldwide: Causes, challenges and responses. *Nat. Med.* **2004**, *10*, S122–S129. [CrossRef] [PubMed]
7. Honda, T.; Uehara, T.; Matsumoto, G.; Arai, S.; Sugano, M. Neutrophil left shift and white blood cell count as markers of bacterial infection. *Clin. Chim. Acta* **2016**, *457*, 46–53. [CrossRef] [PubMed]
8. Austin, B. The value of cultures to modern microbiology. *Antonie Van Leeuwenhoek* **2017**, *110*, 1247–1256. [CrossRef]
9. Gadsby, N.J.; Russell, C.D.; McHugh, M.P.; Mark, H.; Conway, M.A.; Laurenson, I.F.; Hill, A.T.; Templeton, K.E. Comprehensive Molecular Testing for Respiratory Pathogens in Community—Acquired Pneumonia. *Clin. Infect. Dis.* **2016**, *62*, 817–823. [CrossRef] [PubMed]
10. World Health Organization. WHO Coronavirus (COVID-19) Dashboard. Available online: https://covid19.who.int/ (accessed on 12 October 2021).
11. Aziz, A.; Asif, M.; Ashraf, G.; Yang, Q.; Wang, S. COVID-19 Impacts, Diagnosis and Possible Therapeutic Techniques: A Comprehensive Review. *Curr. Pharm. Des.* **2021**, *27*, 1170–1184. [CrossRef] [PubMed]
12. Mishra, N.; Ng, J.; Rakeman, J.L.; Perry, M.J.; Centurioni, D.A.; Dean, A.B.; Price, A.; Thakkar, R.; Angus, A.G.; Williamson, P.; et al. One-step pentaplex real-time polymerase chain reaction assay for detection of zika, dengue, chikungunya, West nile viruses and a human housekeeping gene. *J. Clin. Virol.* **2019**, *120*, 44–50 [CrossRef] [PubMed]
13. Lee, S.H.; Ruan, S.Y.; Pan, S.C.; Lee, T.F.; Chien, J.Y.; Hsueh, P.R. Performance of a multiplex PCR pneumonia panel for the identification of respiratory pathogens and the main determinants of resistance from the lower respiratory tract specimens of adult patients in intensive care units. *J. Microbiol. Immunol. Infect.* **2019**, *52*, 920–928. [CrossRef] [PubMed]
14. Hawkins, S.F.C.; Guest, P.C. Multiplex Analyses Using Real-Time Quantitative PCR. *Mult. Biomark. Tech.* **2017**, *1546*, 125–133.
15. Notomi, T.; Okayama, H.; Masubuchi, H.; Yonekawa, T.; Watanabe, K.; Amino, N.; Hase, T. Loop-mediated isothermal amplification of DNA. *Nucleic Acids Res.* **2000**, *28*, e63. [CrossRef]
16. Piepenburg, O.; Williams, C.H.; Stemple, D.L.; Armes, N.A. DNA detection using recombination proteins. *PLoS Biol.* **2006**, *4*, e204. [CrossRef] [PubMed]
17. Compton, J. Nucleic acid sequence-based amplification. *Nature* **1991**, *350*, 91–92. [CrossRef] [PubMed]
18. Mayboroda, O.; Katakis, I.; O'Sullivan, C.K. Multiplexed isothermal nucleic acid amplification. *Anal. Biochem.* **2018**, *545*, 20–30. [CrossRef] [PubMed]
19. Srividya, A.; Maiti, B.; Chakraborty, A.; Chakraborty, G. Loop Mediated Isothermal Amplification: A Promising Tool for Screening Genetic Mutations. *Mol. Diagn. Ther.* **2019**, *23*, 723–733. [CrossRef] [PubMed]
20. Jennifer, A.; Delaney, R.M.; Ulrich, J.H.P. Detection of the toxic marine diatom Pseudo-nitzschia multiseries using the RuBisCO small subunit (rbcS) gene in two real-time RNA amplification formats. *Harmful Algae* **2011**, *11*, 54–64.
21. Lobato, I.M.; O'Sullivan, C.K. Recombinase polymerase amplification: Basics, applications and recent advances. *Trac Trends Anal. Chem.* **2018**, *98*, 19–35. [CrossRef]
22. Crannell, Z.; Castellanos-Gonzalez, A.; Nair, G.; Mejia, R.; White, A.C.; Richards-Kortum, R. Multiplexed recombinase polymerase amplification assay to detect intestinal protozoa. *Anal. Chem.* **2016**, *88*, 1610–1616. [CrossRef] [PubMed]
23. Larrea-Sarmiento, A.; Stack, J.P.; Alvarez, A.M.; Arif, M. Multiplex recombinase polymerase amplification assay developed using unique genomic regions for rapid on-site detection of genus Clavibacter and C. nebraskensis. *Sci. Rep.* **2021**, *11*, 12017. [CrossRef] [PubMed]
24. Tanner, N.A.; Zhang, Y.; Evans, T.C. Simultaneous multiple target detection in real-time loop-mediated isothermal amplification. *Biotechniques* **2012**, *53*, 81–89. [CrossRef]
25. Lin, W.H.; Zou, B.J.; Song, Q.X.; Zhou, G.H. Progress in multiplex loop-mediated isothermal amplification technology. *Yi Chuan Hered.* **2015**, *37*, 899–910.
26. Curtis, K.A.; Morrison, D.; Rudolph, D.L.; Shankar, A.; Bloomfield, L.S.; Switzer, W.M.; Owen, S.M. A multiplexed RT-LAMP assay for detection of group M HIV-1 in plasma or whole blood. *J. Virol. Methods* **2018**, *255*, 91–97. [CrossRef] [PubMed]
27. Kim, J.H.; Kang, M.; Park, E.; Chung, D.R.; Kim, J.; Hwang, E.S. A Simple and Multiplex Loop-Mediated Isothermal Amplification (LAMP) Assay for Rapid Detection of SARS-CoV. *Biochip J.* **2019**, *13*, 341–351. [CrossRef] [PubMed]
28. Hashimoto, K.; Inada, M.; Ito, K. Multiplex Real-Time Loop-Mediated Isothermal Amplification Using an Electrochemical DNA Chip Consisting of a Single Liquid-Flow Channel. *Anal. Chem.* **2019**, *91*, 3227–3232. [CrossRef]
29. Ma, Y.D.; Luo, K.; Chang, W.H.; Lee, G.B. A microfluidic chip capable of generating and trapping emulsion droplets for digital loop-mediated isothermal amplification analysis. *Lab Chip* **2018**, *18*, 296–303. [CrossRef] [PubMed]
30. Rane, T.D.; Chen, L.; Zec, H.C.; Wang, T.H. Microfluidic continuous flow digital loop-mediated isothermal amplification (LAMP). *Lab Chip* **2015**, *15*, 776–782. [CrossRef]
31. Kalsi, S.; Valiadi, M.; Tsaloglou, M.-N.; Parry-Jones, L.; Jacobs, A.; Watson, R.; Turner, C.; Amos, R.; Hadwen, B.; Buse, J.; et al. Rapid and sensitive detection of antibiotic resistance on a programmable digital microfluidic platform. *Lab Chip* **2015**, *15*, 3065–3075. [CrossRef]
32. Kim, Y.; Yaseen, A.B.; Kishi, J.Y.; Hong, F.; Saka, S.K.; Sheng, K.; Gopalkrishnan, N.; Schaus, T.E.; Yin, P. Single-strand RPA for rapid and sensitive detection of SARS-CoV-2 RNA. *medRxiv* **2020**. [CrossRef]

33. Zhang, Y.; Chen, M.; Liu, C.; Chen, J.; Luo, X.; Xue, Y.; Liang, Q.; Zhou, L.; Tao, Y.; Li, M.; et al. Sensitive and rapid on-site detection of SARS-CoV-2 using a gold nanoparticle-based high-throughput platform coupled with CRISPR/Cas12-assisted RT-LAMP. *Sens. Actuators B* **2021**, *345*, 130411. [CrossRef] [PubMed]
34. Song, J.; Liu, C.; Mauk, M.G.; Rankin, S.C.; Lok, J.B.; Greenberg, R.M.; Bau, H.H. Two-Stage Isothermal Enzymatic Amplification for Concurrent Multiplex Molecular Detection. *Clin. Chem.* **2017**, *63*, 714–722. [CrossRef] [PubMed]
35. El-Tholoth, M.; Bau, H.H.; Song, J. A Single and Two-Stage, Closed-Tube, Molecular Test for the 2019 Novel Coronavirus (COVID-19) at Home, Clinic, and Points of Entry. *ChemRxiv* **2020**. [CrossRef]
36. Huang, G.; Wang, C.; Ma, L.; Yang, X.; Yang, X.; Wang, G. Sensitive sequence-specific molecular identification system comprising an aluminum micro-nanofluidic chip and associated real-time confocal detector. *Anal. Chim. Acta.* **2011**, *695*, 1–10. [CrossRef] [PubMed]
37. Huang, G.; Huang, Q.; Xie, L.; Xiang, G.; Wang, L.; Xu, H.; Ma, L.; Luo, X.; Xin, J.; Zhou, X.; et al. A rapid, low-cost, and microfluidic chip-based system for parallel identification of multiple pathogens related to clinical pneumonia. *Sci Rep.* **2017**, *7*, 6441. [CrossRef] [PubMed]
38. Lin, X.; Jin, X.; Xu, B.; Wang, R.; Fu, R.; Su, Y.; Jiang, K.; Yang, H.; Lu, Y.; Guo, Y.; et al. Fast and Parallel Detection of Four Ebola Virus Species on a Microfluidic-Chip-Based Portable Reverse Transcription Loop-Mediated Isothermal Amplification System. *Micromachines* **2019**, *10*, 777. [CrossRef] [PubMed]
39. Wu, H.; Zhao, P.; Yang, X.; Li, J.; Zhang, J.; Zhang, X.; Zeng, Z.; Dong, J.; Gao, S.; Lu, C. A Recombinase Polymerase Amplification and Lateral Flow Strip Combined Method That Detects Salmonella enterica Serotype Typhimurium with No Worry of Primer-Dependent Artifacts. *Front. Microbiol.* **2020**, *11*, 1015. [CrossRef]

micromachines

MDPI

Article

Numerical Simulation of the Photobleaching Process in Laser-Induced Fluorescence Photobleaching Anemometer

Yu Chen, Shuangshuang Meng, Kaige Wang, Jintao Bai and Wei Zhao *

State Key Laboratory of Photon-Technology in Western China Energy, International Collaborative Center on Photoelectric Technology and Nano Functional Materials, Institute of Photonics & Photon Technology, Northwest University, Xi'an 710127, China; 201920590@stumail.nwu.edu.cn (Y.C.); 201931692@stumail.nwu.edu.cn (S.M.); wangkg@nwu.edu.cn (K.W.); baijt@nwu.edu.cn (J.B.)

* Correspondence: zwbayern@nwu.edu.cn

Abstract: At present, a novel flow diagnostic technique for micro/nanofluidics velocity measurement—laser-induced fluorescence photobleaching anemometer (LIFPA)—has been developed and successfully applied in broad areas, e.g., electrokinetic turbulence in micromixers and AC electroosmotic flow. Nevertheless, in previous investigations, to qualitatively reveal the dynamics of the photobleaching process of LIFPA, an approximation of uniform laser distribution was applied. This is different from the actual condition where the laser power density distribution is normally Gaussian. In this investigation, we numerically studied the photobleaching process of fluorescent dye in the laser focus region, according to the convection–diffusion reaction equation. The profiles of effective dye concentration and fluorescence were elucidated. The relationship between the commonly used photobleaching time constant obtained by experiments and the photochemical reaction coefficient is revealed. With the established model, we further discuss the effective spatial resolution of LIFPA and study the influence of the detection region of fluorescence on the performance of the LIFPA system. It is found that at sufficiently high excitation laser power density, LIFPA can even achieve a super-resolution that breaks the limit of optical diffraction. We hope the current investigation can reveal the photobleaching process of fluorescent dye under high laser power density illumination, to enhance our understanding of fluorescent dynamics and photochemistry and develop more powerful photobleaching-related flow diagnostic techniques.

Keywords: microfluidics; photobleaching process; laser-induced fluorescence photobleaching anemometer; super-resolution

Citation: Chen, Y.; Meng, S.; Wang, K.; Bai, J.; Zhao, W. Numerical Simulation of the Photobleaching Process in Laser-Induced Fluorescence Photobleaching Anemometer. *Micromachines* **2021**, *12*, 1592. https://doi.org/10.3390/mi12121592

Academic Editor: Aiqun Liu

Received: 8 November 2021
Accepted: 17 December 2021
Published: 20 December 2021

1. Introduction

Laser-induced fluorescence photobleaching anemometer (LIFPA) is a velocity measurement technique developed by Wang in 2005 [1] and primarily for micro/nanofluidics. It measures flow velocity due to the photobleaching nature of fluorescent dye under a high laser power density. When fluorescent dye molecules are excited by a wavelength-matched laser, either the structures of molecules could be damaged due to the photo instability of the dye molecules [2], or the excited states could be quenched by dye–dye or solvent interactions, when the power density of the excitation beam exceeds a certain threshold. The photobleaching results in a decrease in the effective concentration (C) of fluorescent dye and a weakening fluorescent signal with increasing irradiation time [3]. The faster the fluorescent molecules pass through the excitation region in a flow field with uniform fluorescent molecules, the higher the effective concentration of fluorescent dye, and the higher the fluorescent intensity collected [4]. By establishing the monotonic relationship [5] between fluorescence intensity and flow velocity, the velocity of the flow field can be calculated by detecting the fluorescence signal intensity in the spot area [6]. This technique has been successfully applied to the velocity measurement of complex flow fields such as linearly and nonlinearly oscillating electroosmotic flow [7–10], and microelectrokinetic turbulence [11–15].

The dynamics of photobleaching is the kernel of LIFPA applications and determines the performance of the LIFPA system. Wang qualitatively analyzed the photobleaching process [1], attributing it to a first-order model of reaction:

$$I_f = I_{f0} \times \exp(-t/\tau) = I_{f0} \times \exp\left(-d_f/U\tau\right) \tag{1}$$

where I_f represents the overall fluorescence intensity of the laser focal area, I_{f0} is the initial I_f at $t = 0$, t is the bleaching time that the fluorescent molecules pass through the region of laser focus diameter d_f, τ is the photobleaching time constant, U represents flow velocity. Then, Zhao et al. [16] studied the photobleaching process and established a theoretical formula to estimate the temporal resolution of LIFPA.

$$I_f = d_f I_{f0} \Psi_s \, U\tau \left(1 - e^{-\frac{d_f}{U\tau}}\right) + I_{f,b} \tag{2}$$

where Ψ_s is the weight function of U, and $I_{f,b}$ is the background fluorescence intensity when the dye is approximately fully photobleached. However, to simplify the analysis, they assumed the local fluorescence intensity ($I_{f,local}$) distribution in the focal area to be uniform, which could lead to deviation from the actual case, where $I_{f,local}$ normally has a Gaussian distribution. Moreover, the relationships among the complex fluorescent dynamics, the photobleaching reaction process and the superficial and measurable experimental parameters (e.g., τ) have not been established. This could be a major obstacle to the development of the LIFPA technique.

In this investigation, the photobleaching process is studied as a photochemical reaction, which can be further numerically analyzed by a convection–diffusion reaction process. We first established a relationship between the photobleaching time constant and photochemical reaction coefficient, by comparing experiments and numerical simulations, to reveal the relationships between the photobleaching reaction process and experimental parameters. Then, the influence of flow velocity and laser intensity on the effective concentration and fluorescent intensity was numerically investigated. At last, the spatial resolution of LIFPA was revisited to better understand the photobleaching phenomenon in flowing liquids and to develop new flow diagnostic techniques.

2. LIFPA Photobleaching Model

The photobleaching process of the fluorescent solution in a steady microchannel flow can be described by a convection–diffusion reaction equation [17–19] as

$$\frac{\partial C}{\partial t} + \vec{u} \cdot \nabla C = D\nabla^2 C + R_c \tag{3}$$

In Equation (3), C represents the concentration of the fluorescent solution, $\vec{u}$ is the flow velocity vector, D is the diffusivity of the dye molecule in the solution, and R_c represents the reaction term according to photobleaching, which is determined by the laser intensity distribution, C and a factor k_b. By first-order approximation, R_c can be expressed as

$$R_c = -k_b I C \tag{4}$$

where k_b represents the photobleaching rate of dye molecules per unit excitation intensity at a specific excitation wavelength, with a unit of $\text{cm}^2/(\text{W}\cdot\text{s})$. It represents how fast the fluorescent dye can be photobleached. The higher the k_b, the faster the photobleaching is. I is the intensity of the excitation laser; in general, I has a Gaussian distribution, shown as

$$I = I_{max} G(x, y, z) \tag{5}$$

I_{max} is the peak value of I, and $G(x,y,z)$ is a 3D Gaussian function, shown as

$$G(x,y,z) = \exp\left[-\left(\frac{(x-x_0)^2}{2\sigma_x{}^2} + \frac{(y-y_0)^2}{2\sigma_y{}^2} + \frac{(z-z_0)^2}{2\sigma_z{}^2}\right)\right] \quad (6)$$

where σ_x, σ_y and σ_z represent the standard deviations of laser intensity in x, y and z directions, respectively. Substituting Equations (4)–(6) into Equation (3), with dimensional analysis by

$$C = C^*C_0, x = x^*L, y = y^*L, z = z^*L, \vec{u} = \vec{u}^*U_0, \nabla = \nabla^*/L, t = t^*L/U_0 \quad (7)$$

where C_0, L and U_0 are characteristic concentration, length scale and velocity, respectively, we have

$$\frac{\partial C^*}{\partial t^*} + \vec{u}^* \cdot \nabla^* C^* = \frac{1}{P_e}\nabla^{*2}C^* - \frac{k_b I_{\max} L}{U_0} G(x^*,y^*,z^*)C^* \quad (8)$$

where $P_e = U_0L/D$ is the Péclet number which represents the ratio of the convection rate to the diffusion rate. In microfluidic applications, normally the disturbance of microflow has the smallest time scale of O(10^{-4} s) or higher. In contrast, to obtain good sensitivity and resolution, LIFPA has a fast photobleaching process where τ is of O(10^{-6} s), which is approaching the lower limit of photobleaching lifetime, which is normally in the range of nanoseconds to seconds [20–24], or even longer. Meanwhile, the resident time of dye in the laser focus is O(10^{-4} s) or lower. Therefore, the time scale of flow is equivalent to or larger than the resident time and τ. A quasi-steady state of the photobleaching process of dye can be approximated in the laser focus region. Thus, $\partial C^*/\partial t^* = 0$. The LIFPA photobleaching model under a quasi-steady state is obtained as

$$\vec{u}^* \cdot \nabla^* C^* = \frac{1}{P_e}\nabla^{*2}C^* - Z_{pc}C^*G(x^*,y^*,z^*) \quad (9)$$

where $Z_{pc} = KL/U_0$ is a dimensionless number to evaluate the ratio between the photobleaching rate and convective transport rate of dye molecules. The larger the Z_{pc}, the stronger the local photobleaching is. Otherwise, the convective transport of the concentration of the fluorescent solution is more important. $K = k_b I_{\max}$ is the photochemical reaction coefficient, which represents the photobleaching rate of dye molecules in this investigation. It is a crucial parameter to characterize the photobleaching process through the convection–diffusion reaction equation in COMSOL simulation. Although K has a unit of 1/s, which is the same as $1/\tau$, K is not $1/\tau$. One of the focuses of this investigation is to find the relationship between K and τ. The characteristic concentration C_0 was eliminated during the simplification process. This indicates that increasing or decreasing the dye concentration C does not affect the first-order photobleaching process, if in the absence of light absorption. Finally, the local fluorescent intensity $I_{f,local}$ can be calculated accordingly as [25]

$$I_{f,local} = q_f IC \quad (10)$$

with q_f being a fluorescence quantum yield factor.

3. Experimental Setup

3.1. LIFPA System

The experimental LIFPA system is developed based on a confocal microscope, as shown in Figure 1a. The excitation laser is a 405 nm continuous wave laser (MDL-III-405-500, CNI). The excitation beam is first controlled by an acousto-optic modulator (AOM, 1206C-2-1002, Isomet) for on–off switch. Then, it passes through a spatial pinhole filter (SLF, SFB-16DM, OptoSigma) to improve the beam quality. A diaphragm is used to filter out the zero-order and high-order diffraction spots and retain the first-order spot in the center.

After passing through a collimation lens, a collimated excitation beam with a Gaussian beam profile is obtained.

Subsequently, the excitation beam passes through the dichroic mirror (DM_1), which transmits the 405 nm laser and reflects the fluorescence around 480 nm. The excitation beam is then reflected into the objective lens (OL, Olympus PlanApo, 100 × NA 1.4 oil immersion Objective lens) by a mirror (M_1). The excitation beam is finally focused in the microchannel flow and illuminates the fluorescent dye. The fluorescent signal passes the optical path along M_1, DM_1, and a second mirror (M_2), and then passes through a band-pass filter (BP, 470/10 nm, OptoSigma) to eliminate background light noise. The fluorescent signal is collected by a lens (L_2) and focused on a pinhole PH_1 for spatial filtering. The filtered fluorescent signal is detected by a photon counter (H7421-40, Hamamatsu). The exposure time of fluorescent measurement, i.e., the sampling duration, is 0.5 ms. A time interval of 0.5 ms is applied between two samples. Therefore, the sampling rate is 1 kHz. Since the photo counter has a saturation count rate of 1.5×10^6 photons per second for linear measurement (equivalent to a maximum of 750 photons in the 0.5 ms sampling duration), and the fluorescence of LIFPA is not weak, we used a neutral density filter of OD2 in front of PH_1, to reduce the fluorescent intensity and guarantee there is no nonlinear saturation of photon counter. In the experiments, the maximum photon number in a 0.5 ms sampling duration is below 533, which is in the linear range of the photon counter.

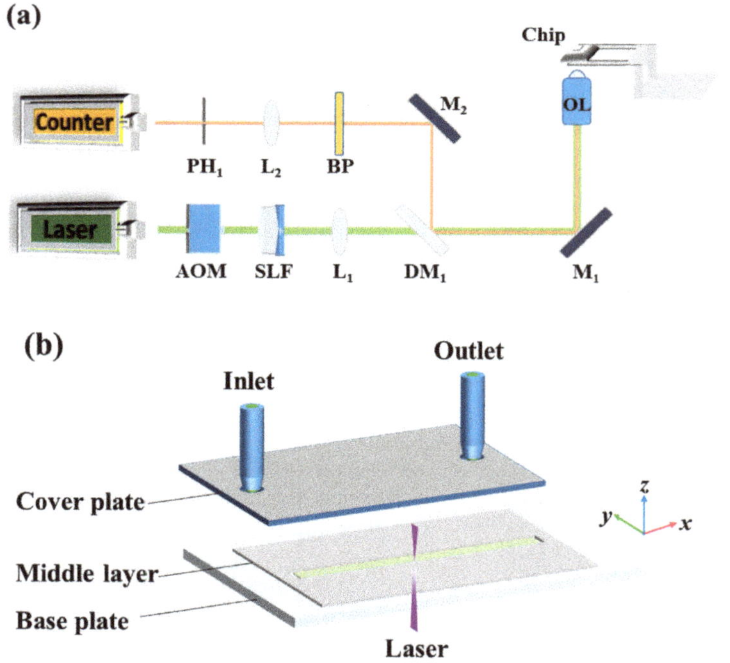

Figure 1. Experimental setup. (**a**) Schematic diagram of LIFPA experimental system based on confocal microscope; (**b**) schematic diagram of microchip.

3.2. Microchannel and Solution Preparation

The experiment is carried out at the center of the microchannel. For precise positioning of the microchannel with a large travel distance, a 2D translation stage (PI M-521.DG, 1 µm accuracy) and a high-precision 3D piezo nanostage (PI 562.3CD, 1 nm accuracy) are applied together, as shown in Figure 1a. The structure of the microfluidic chip is shown in Figure 1b. It has three layers. The cover layer is made of 2 mm-thick acrylic with good transparency and mechanical strength. The middle layer, which is also the channel layer, is made of plastic sheets. The bottom plate is a low-fluorescence glass slide with a thickness of 130 μm. The chip is assembled layer-by-layer, to realize a microchannel with a rectangular cross section. The length, width, and height of the microchannel are 5 mm, 360 μm, and 90 μm, respectively. During experiments, the fluorescent solution of Coumarin 102 (C102, Sigma Aldrich) at a concentration of 0.1 mM is injected into the microchannel with a syringe pump to generate a laminar flow. The C102 solution is prepared by dissolving 2.5 mg of C102 powder with 95 mL of deionized water and 5 mL of methanol solution (Analytical Reagent, concentration > 99.5%). The C102 solution has a pH value of 7.82. Finally, after excitation, the photobleached fluorescent solution is drained from the outlet.

4. Numerical Simulation and Experiment

4.1. Numerical Simulation by COMSOL

In this paper, the photobleaching process of LIFPA is numerically investigated with COMSOL Multiphysics 4.3. The simulation is carried out in a 3D region with rectangular cross sections, as shown in Figure 2a.

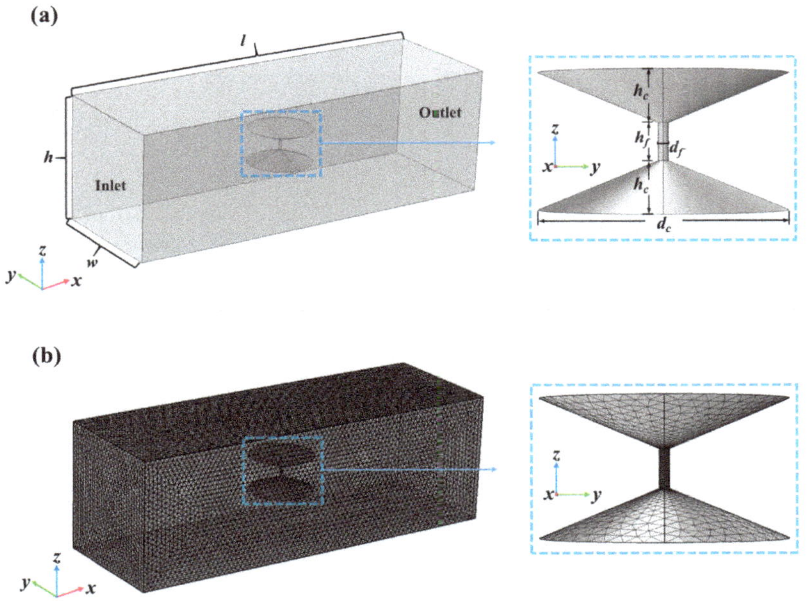

Figure 2. LIFPA simulation model in COMSOL. The coordinate origin is located at the geometric center. (**a**) Geometry of computation model of LIFPA. The inset shows the structural details of the laser focus region, which is also the reaction region; (**b**) details of mesh in COMSOL simulation.

A laser focus region approximately simulating the laser beam is located at the center ($x = y = z = 0$) of the computational region and has an axisymmetric structure with 3 parts. At the middle part is a cylindrical region with diameter $d_f = 203$ nm and height $h_f = 800$ nm, whose diameter is coincident with the full width at half maximum (FWHM) of the laser beam. The top and bottom parts are symmetric cones with diameter $d_c \approx 5$ μm and height $h_c = 1$ μm, respectively, to be coincident with the light field according to the high NA of the objective. In the 3D simulation, free tetrahedral elements with variable sizes are applied, as shown in Figure 2b. Since the size of the laser focus region of the excitation beam is only 203 nm, to have sufficient spatial resolution and smooth distribution of concentration field, the minimum size of the elements in the region is 0.5 nm. In the other regions, the minimum and maximum sizes of the elements are 100 nm and 530 nm, respectively. If we simulated the convection–diffusion reaction equation in the entire experimental microchannel, the computation burden is extraordinarily heavy. To save computation resources, in the simulation, we reduce the size of the computation region to 30 μm long, 10 μm wide and 10 μm high. The total number of free tetrahedral elements is 1,311,608.

The photobleaching process is simulated with laminar flow and dilute matter transfer modules in COMSOL. In the laminar flow module, the flow is assumed to be steady, unidirectional and fully developed. Therefore, we have $\vec{u} = U\hat{x}$ for simplification. A no-slip boundary condition [26,27] is been applied on the side walls (Ω) as

$$U|_{\Omega} = 0 \tag{11}$$

Since the sizes of the experimental microchannel are different from those of the computational region, to guarantee that the flow velocities at the laser focus regions of both experiments and simulations are the same, the flow rate in numerical simulation is different from that in experiments. In this investigation, we first calculate the flow velocity at the laser focus region (the third row of Table 1) through experimental flow rate (the first row of Table 1), according to Equation (12) [8,28,29]:

$$Uy,z = \frac{48Q}{\pi^3 wh}\left\{\sum_{n,odd}^{\infty}\frac{1}{n^3}\left[1-\frac{\cos h\left(\frac{n\pi y}{h}\right)}{\cos h\left(\frac{n\pi w}{2h}\right)}\right]\sin\left(\frac{n\pi z}{h}\right)\right\}\left[1-\sum_{n,odd}^{\infty}\frac{192h}{n^5\pi^5 w}\tan h\left(\frac{n\pi w}{2h}\right)\right]^{-1} \tag{12}$$

where Q is the flow rate. $-w/2 \leq y \leq w/2$ and $-h/2 \leq z \leq h/2$. Then, the inlet flow rate in the simulation (the second row of Table 1) is calculated based on the flow velocity at the laser focus region (the third row of Table 1) in simulation, which is the same as in experiments. The outlet pressure is the circumstantial pressure. The flow is incompressible with a constant temperature.

In the dilute matter transfer module, the diffusion coefficient of molecules in the solution is $D = 1 \times 10^{-9}$ m^2/s. The initial fluorescent solution concentration $C = 0.1$ mM. At the side walls, no flux is present. The photobleaching of fluorescent molecules occurs only in the laser focus region in Figure 2. The photochemical reaction is stimulated by applying a reaction term based on Equations (4)–(6).

Table 1. Flow conditions in experiments and numerical simulations.

Flow Rate in experiments (μL/min)	5.0	10.0	15.0	20.0	25.0	30.0	35.0	40.0	45.0
Flow rate in simulation (nL/min)	13.1	26.1	39.2	52.2	65.3	78.3	91.4	104.5	117.5
U(0,0) for both experiments and simulations (mm/s)	4.6	9.1	13.7	18.2	22.8	27.4	31.9	36.5	41.1

4.2. Direct Comparison between Experiments and Numerical Simulations

The relationships between U and fluorescence are established by experiments at three different laser powers (P) of 6.9, 11.8 and 18.2 mW. The corresponding $I_{\max}$ are 2.96 × 10^7, 5.05 × 10^7 and 7.80 × 10^7 W/cm^2, respectively, as shown in Figure 3a. Each data point was calculated by averaging over 2 × 10^4 fluorescent signals, i.e., a 20 s time sequence of fluorescent signals. It is obvious that I_f increases with the increasing U. Moreover, when the laser power density is small (e.g., $I_{\max} = 2.96 \times 10^7$ W/cm^2), I_f shows an obvious nonlinear increment with U. When $I_{\max}$ is increased to 5.05 × 10^7 and 7.80 × 10^7 W/cm^2, the nonlinear behavior gradually decreases and the curve becomes more linear, accompanied by a decreasing slope of the $U \sim \widetilde{I}_f$ curve. In these cases, velocity fluctuations can be more clearly distinguished by larger fluorescent fluctuations, as shown in Figure 3a. Thus, high laser power density can provide a better sensitivity of LIFPA measurement.

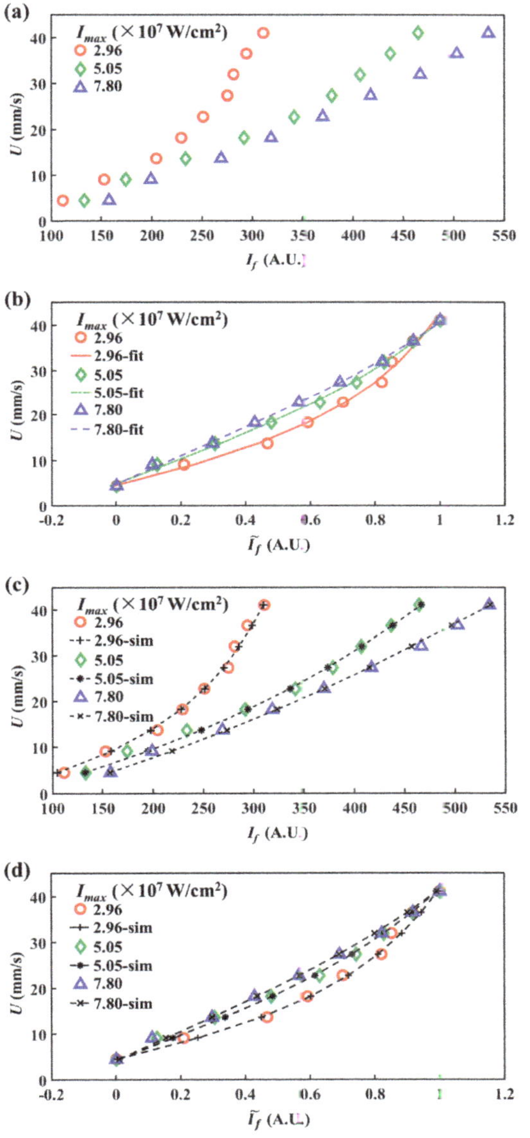

Figure 3. LIFPA velocity calibration curves. 'fit' and 'sim' represent the fitting and simulation calibration curves, respectively. (**a**) Experimental calibration curves for different laser power densities at center of the microchannel; (**b**) normalization for (**a**); (**c**) comparison of experimental and simulated calibration curves; (**d**) normalization for (**c**).

In order to compare $U \sim \widetilde{I_f}$ relationships at different laser power densities, we calculate the normalized fluorescence intensity ($\widetilde{I_f}$) with the maximum ($I_{f,\max}$) and minimum ($I_{f,\min}$) values as [2]

$$\widetilde{I_f} = \frac{I_f - I_{f,\min}}{I_{f,\max} - I_{f,\min}} \tag{13}$$

The $U \sim \widetilde{I_f}$ curves are shown in Figure 3b. It can be clearly seen that the lower the laser power density of the excitation laser, the stronger the curve bending. After normalization, the $U \sim \widetilde{I_f}$ curves under each $I_{\max}$ only lie in the different bending. In the experiments, we use the photobleaching time constant (τ) to evaluate the strength of the photobleaching. τ can be approximately calculated by nonlinear fitting on $U \sim \widetilde{I_f}$ through the equation below [2]

$$\widetilde{I_f} = aU\left[1 - \exp\left(-\frac{d_f}{U\tau}\right)\right] + b \tag{14}$$

where a is an amplification factor which dominates the slope of the $U \sim \widetilde{I_f}$ curve and b represents the initial fluorescence intensity of I_f at $U = 0$. τ determines the curvature of the curve. According to Equation (14), a set of $U \sim \widetilde{I_f}$ curves are nonlinearly fitted (as shown in Figure 3b) to calculate the corresponding τ. The results of τ calculated from the experiments are listed in Table 2. It can be seen, as P is increased, that τ apparently decreases to as low as 2.43 μs.

Subsequently, the numerically calculated $U \sim \widetilde{I_f}$ curve is compared with that of the experiments. In the simulation, $I_f = \int\int\int_{\mathbb{D}} I_{f,\text{local}} dV$, where $\mathbb{D}$ is a cylindrical integration region (i.e., detection region) for fluorescent collection, with a height of h_f and diameter of d_{cl}. In this section, $\mathbb{D}$ is the cylindrical part of the inset of Figure 2a and $d_{cl} = d_f$, to be coincident with the experimental system we applied. As shown in Figure 3c,d, both the original and normalized numerically calculated fluorescence curves show high consistency with the experimental ones. Since the numerical calculated $U \sim \widetilde{I_f}$ curves are merely determined by the photochemical reaction coefficient K, accordingly, a relationship between τ and K can be established, as shown in Figure 4 and Table 2. When $I_{\max}$ is increased, we find K increases with $I_{\max}$ in an approximately linear manner, while K decreases with τ as $K \sim \tau^{-2.28}$, as shown in the inset of Figure 4.

This indicates that the fluorescent photobleaching time shows a nonlinear relationship with K under strong laser illumination. To achieve a higher temporal resolution with smaller τ, K should be significantly increased.

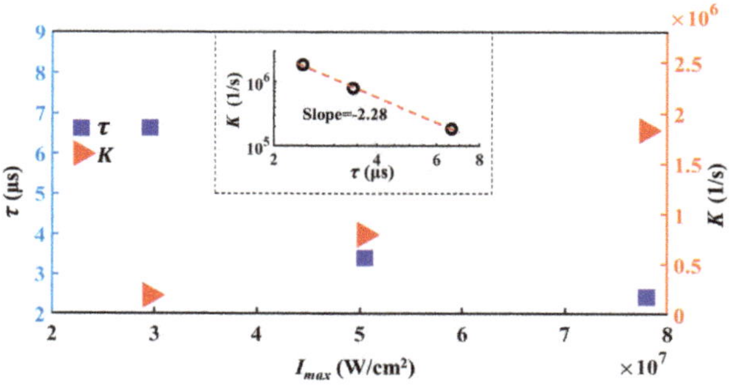

Figure 4. Comparisons among photochemical reaction coefficient K, photobleaching time constant τ and laser power density I_{max}. The inset displays the relationship of K and τ in a log–log plot.

Table 2. Experimental results of τ, peak I (say I_{max}) and the corresponding photochemical reaction coefficient K in numerical simulation under different P.

P (mW)	6.90	1.80	18.20
I_{max} (W/cm^2)	2.96×10^7	5.05×10^7	7.80×10^7
τ (µs)	6.63	3.41	2.43
K (1/s)	1.85×10^5	7.95×10^5	1.83×10^6

4.3. Effective Concentration Distribution

The influence of photobleaching on the effective concentration C of fluorescent dye can be directly observed in Figure 5. The photobleaching generates a cometlike region of C. When the flow velocity U at the focus region is small (e.g., $U = 4.6$ mm/s), as shown in Figure 5a, a strongly photobleached region can be found in the focus region of the laser. The cometlike region of C has a large width attributed to the lateral diffusion. When U is increased, the fluorescent dye is less photobleached and C in the focus region is significantly higher than the counterparts of smaller U, as can be seen from Figure 5b,c. The tails of the cometlike regions exhibit larger length and smaller width. Downstream of the focus area, C gradually recovers due to the molecular diffusion. Similar results can also be found in the second and third rows of Figure 5.

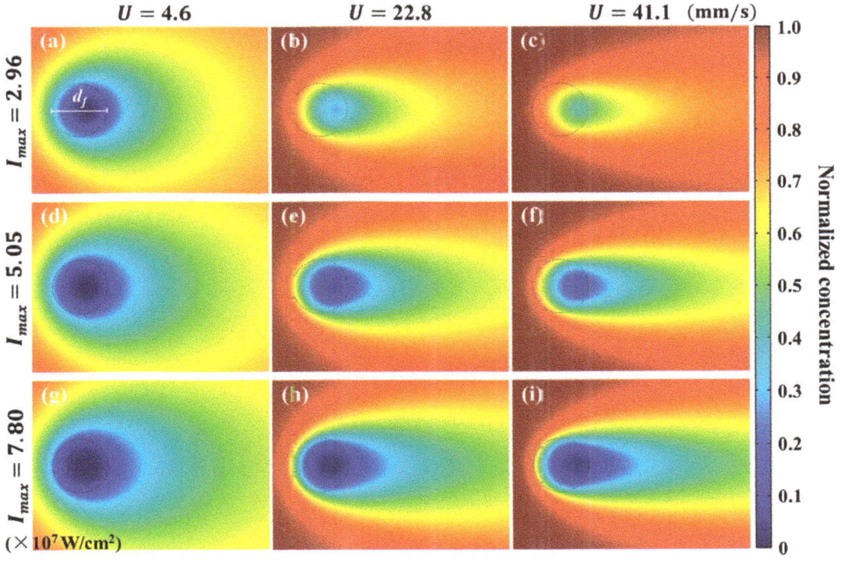

Figure 5. Distribution of effective concentration C of fluorescent dye under different bulk flow velocities and laser power densities. **(a–c)** C under $I_{max} = 2.96 \times 10^7$ W/cm^2, with $U = 4.6$, 22.8, 41.1 mm/s, respectively. **(d–f)** C under $I_{max} = 5.05 \times 10^7$ W/cm^2, with $U = 4.6, 22.8, 41.1$ mm/s, respectively. **(g–i)** under $I_{max} = 7.80 \times 10^7$ W/cm^2, with $U = 4.6$, 22.8, 41.1 mm/s, respectively. Here, the concentration values have been normalized by the maximum C. The black circle represents the integration region of fluorescence to calculate I_f.

The distribution of C is dominated by three dimensionless numbers, i.e., P_e and Z_{pc}, in Equation (9), and $Z_{pd} = Z_{pc}P_e$, which is the ratio between photobleaching rate and diffusion rate. When P_e is increased with constant Z_{pc}, the influence of diffusivity becomes smaller and the width of the cometlike region is decreased. When Z_{pc} is increased with constant P_e, the influence of photobleaching is enhanced and the value of C becomes smaller in the focus area. When Z_{pd} is fixed, the tail length of the cometlike region of C increases with P_e.

4.4. Fluorescence Intensity Distribution

The distributions of fluorescence under different flow velocities and excited by different laser power densities are calculated according to Equation (10). The results are shown in Figure 6, where the fluorescent intensity is normalized by the maximum value. It is obvious that the fluorescent intensity in the focus area is symmetric on the x axis, and decreases along the flow direction as expected.

As shown in Figure 6a–c, when U is increased with fixed $I_{\max} = 2.96 \times 10^7$ W/cm^2, P_e is increased, while Z_{pc} is decreased. The streamwise positions of the peak fluorescent intensities move downstream and toward the center of the focus area. Similar results can also be found on the second and third rows. When $I_{\max}$ is increased with fixed $U = 41.1$ mm/s, P_e is constant, while Z_{pc} is increased. From Figure 6c,f,i as examples, it can be seen that the streamwise positions of the peak fluorescent intensities move upstream and even leave the integration region. The fluorescent intensity at the center of focus area continuously decreases.

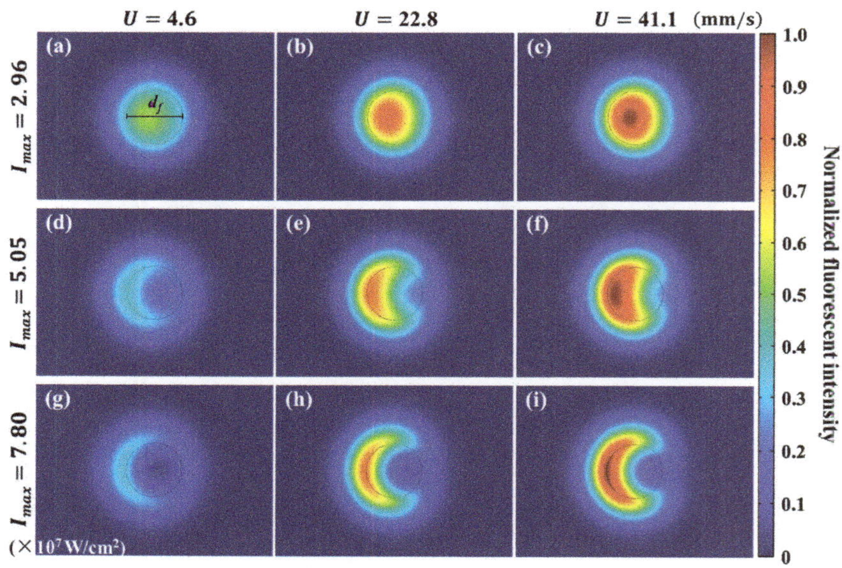

Figure 6. Distribution of fluorescent intensity $I_{f,\text{local}}$ under different bulk flow velocities and laser power densities. **(a–c)** $I_{f,\text{local}}$ under $I_{\max} = 2.96 \times 10^7$ W/cm^2, with $U = 4.6$, 22.8, 41.1 mm/s, respectively. **(d–f)** $I_{f,\text{local}}$ under $I_{\max} = 5.05 \times 10^7$ W/cm^2, with $U = 4.6$, 22.8, 41.1 mm/s, respectively. **(g–i)** $I_{f,\text{local}}$ under $I_{\max} = 7.80 \times 10^7$ W/cm^2, with $U = 4.6$, 22.8, 41.1 mm/s, respectively. Here, the legend is normalized by the peak fluorescent intensity. The black circle represents the integration region of fluorescence to calculate I_f.

5. Velocity Measurement of Breaking Optical Diffraction Limit

Figure 6d,g indicate, as Z_{pc} is increased, the major body of fluorescence moves upstream and even out of the integration region. This could lead to a "waste" of fluorescence for calculating flow velocity. This reminds us to rethink the influence of the integration region on the performance, especially the spatial resolution of the LIFPA system. Therefore, we revisited the spatial resolution of LIFPA first in this section. Then, we discuss a possible method to break the optical diffraction limit and realize a super-resolution velocity measurement.

5.1. Spatial Resolution of Effective Velocity Measurement with LIFPA

To investigate the details of the photobleaching process of LIFPA, we calculated the fluorescent intensity profile $\widehat{I_{f,local}}$ ($\widehat{I_{f,local}} = I_{f,local} / \max\limits_{y=z=0}(I_{f,local})$) through the center of the laser focus spot ($-406 \leq x \leq 406$ nm), as shown in Figure 7. It can be seen, when $K = 1.85 \times 10^5$ 1/s (Figure 7a), the $\widehat{I_{f,local}}$ curves according to different U show a small difference. As K is further increased, as shown in Figure 7b,c, the fluorescent intensity at the center part becomes smaller and the streamwise positions of the peak $\widehat{I_{f,local}}$ move downstream at larger U. One noteworthy phenomenon is that when K is sufficiently large (see Figure 7c,d as an example), there is a valley with flat $\widehat{I_{f,local}}$, where the fluorescent dye molecules have been strongly photobleached. When measuring velocity fluctuations, only the part marked with a width of d_{eff} (i.e., actual resolution, which is defined as the distance from the left edge of the integration region to the position where $\widehat{I_{f,local}} = 0.05\widehat{I_{f,local,1}}$, $\widehat{I_{f,local,1}}$ is marked in Figure 7c and represents the $\widehat{I_{f,local}}$ at the left edge of the integration region) in Figure 7d has an apparent contribution to the fluorescent variation. The other part in the integration region has a negligible contribution to the fluorescent variation, and accordingly, the velocity variation, especially at higher velocity. In other words, the velocity measurement by LIFPA is only sensitive in the marked part, which can be considered as the point spreading function (PSF) of the fluorescent intensity. Therefore, the width of PSF, i.e., d_{eff}, actually determines the spatial resolution of LIFPA.

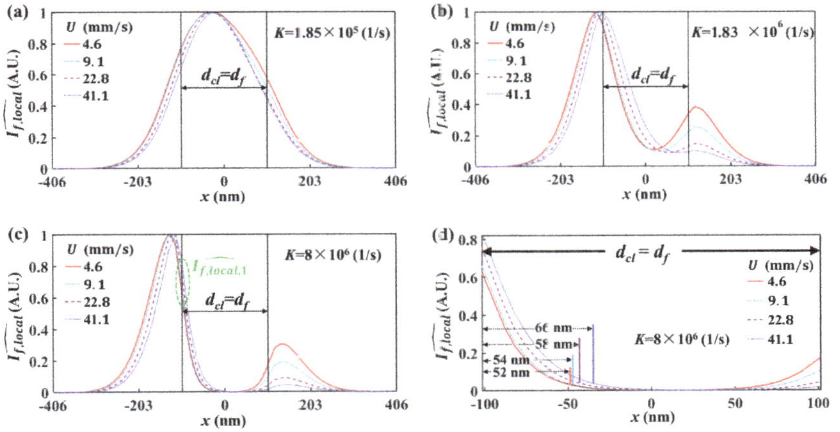

Figure 7. $\widehat{I_{f,local}} \sim x$ curves normalized at different U under different K. Here, $d_{cl} = d_f = 203$ nm. In each figure, the fluorescent intensity profiles are plotted with four different U from 4.6 mm/s to 41.1 mm/s. (**a**) $K = 1.85 \times 10^5$ 1/s, (**b**) $K = 1.83 \times 10^6$ 1/s, (**c**) $K = 8.00 \times 10^6$ 1/s, (**d**) zoomed-in view of the integration region in (**c**).

This is intriguing, since in conventional LIFPA techniques, the spatial resolution of LIFPA is believed to be equivalent to the spatial resolution of the optical system (i.e., the FWHM of the point spread function). From the current research, it can be seen when K is sufficiently large, e.g., $K = 8 \times 10^6$ 1/s, the averaged d_{eff} over the four curves shown in Figure 7d is around 58 nm. The actual spatial resolution of LIFPA can be apparently smaller than that of the optical system (203 nm in this investigation) and the diffraction limit, i.e., the super-resolution of velocity measurement can be realized. Specifically, this can be achieved by simply increasing I_{max} in a confocal microscope, or properly selecting a fluorescent dye with large k_b. For instance, if $K = I_{\max}k_b$ is increased to 2×10^7 1/s, the averaged d_{eff} can theoretically be as low as 30 nm.

Relative to the LIFPA system, developed on the basis of stimulated emission depletion (STED) super-resolution techniques [6] to realize super-resolution velocity measurements, where two beams (one excitation beam and one depletion beam) must be collimated and well aligned, the current investigation shows that LIFPA can be intrinsically super-resolution even with a simple confocal microscope, if K is sufficiently large. Additionally, in the STED LIFPA system, the power density of the excitation beam cannot be very high, otherwise the spatial resolution of STED is reduced. Thus, the temporal resolution of the STED LIFPA system is normally low. However, for a sufficiently large K, both ultrahigh spatial and temporal resolutions can be achieved simultaneously.

5.2. Influence of Integration Region on Velocity Measurement

In this section, we further study the influence of d_{cl} of integration region on the velocity calibration curve, and attempt to show another way to improve LIFPA's spatial resolution. The results are plotted in Figure 8. Relative to Figure 3a where $d_{cl} = d_f$, the velocity calibration curves calculated at $d_{cl} = 0.5d_f$ show clearly smaller values and the value ranges of I_f, as plotted in Figure 8a, in line with our expectations. Although it could lead to a negative influence on velocity measurement due to worse signal-to-noise ratio (SNR), the spatial resolution could be clearly improved to super-resolution level. From Figures 3b and 8b,c, it can be seen that changing d_{cl} will not affect the result that the curves show more bending at lower $I_{\max}$. As d_{cl} is increased, the influence of $I_{\max}$ on the bending of velocity calibration curves becomes smaller, as can be seen from Figure 8b,c. The $U \sim \widetilde{I_f}$ plots become indistinguishable between $I_{\max} = 5.05 \times 10^7$ and 7.80×10^7 W/cm^2 in Figure 8c. We plot four $U \sim \widetilde{I_f}$ curves of different d_{cl} in Figure 8d at $I_{\max} = 7.80 \times 10^7$ W/cm^2. Despite the smaller fluorescent signal at smaller d_{cl}, the velocity calibration curve becomes more linear at smaller d_{cl}. It again indicates the temporal resolution is also improved, accompanied by the spatial resolution.

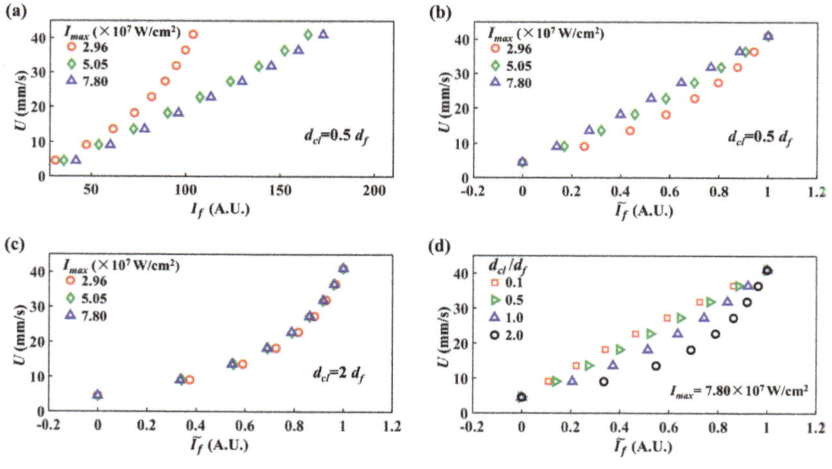

Figure 8. LIFPA velocity calibration curves. (**a**) Original calibration curves for different laser power densities at $d_{cl} = 0.5d_f$; (**b**) normalized calibration curves for (**a**); (**c**) normalized calibration curves for different power densities at $d_{cl} = 2d_f$; (**d**) normalized calibration curves for different d_{cl} at $I_{\max} = 7.80 \times 10^7$ W/cm^2.

In addition, we plotted the $\widehat{I_{f,\text{local}}} \sim x$ under $d_{cl} = 0.5d_f$ in Figure 9. It can be seen that as d_{cl} is reduced, the corresponding d_{eff} is also reduced. At $K = 8.00 \times 10^6$ 1/s, the averaged d_{eff} over the four curves is around 55 nm, which is slightly smaller than that at

$d_{cl} = d_f$ shown in Figure 7d. Thus, by decreasing the size of the integration region, the actual spatial resolution of LIFPA can be further enhanced.

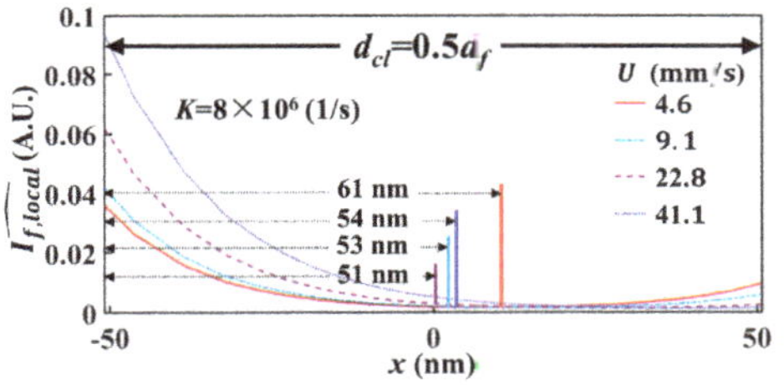

Figure 9. Zoomed-in view of the integration region in the $\widehat{I_{f,\mathrm{local}}} \sim x$ curves normalized at different U, if $d_{cl} = 0.5d_f$ and $K = 8.00 \times 10^6$ 1/s.

6. Conclusions

In this investigation, we studied the photobleaching process of fluorescent dye in the laser focus region, according to the model based on the convection–diffusion reaction equation. The profiles of effective dye concentration and fluorescence were elucidated. The relationship between the commonly used photobleaching time constant obtained by experiments and the photochemical reaction coefficient was revealed. We further studied the influence of the detection region of fluorescence on the performance of the LIFPA system, and found that at sufficiently high excitation laser power density, LIFPA can achieve super-resolution and break the limit of optical diffraction, even in a confocal microscope. We hope the current investigation can promote the development of LIFPA, and reveal the photobleaching process of fluorescent dye under high laser power density illumination, to enhance our understanding of fluorescent dynamics and photochemistry and develop more powerful photobleaching-related flow diagnostic techniques.

Author Contributions: Conceptualization, W.Z.; methodology, Y.C. and S.M.; software, Y.C.; validation, W.Z.; formal analysis, Y.C.; investigation, Y.C.; data curation, Y.C.; writing—original draft preparation, Y.C.; writing—review and editing, K.W., J.B. and W.Z.; supervision, W.Z.; project administration, K.W., J.B. and W.Z.; funding acquisition, K.W., J.B. and W.Z. All authors have read and agreed to the published version of the manuscript.

Funding: This investigation is supported by the National Natural Science Foundation of China (grant no. 11672229, 51927804, 61775181, 61378083), and the Natural Science Basic Research Program of Shaanxi Province—Major Basic Research Project (2018TD-018, 2016ZDJC-15).

Acknowledgments: We thank Yang Dang and Ping Ma for valuable discussions.

Conflicts of Interest: The authors declare no conflict of interest.

References

1. Wang, G.R. Laser induced fluorescence photobleaching anemometer for microfluidic devices. *Lab Chip* **2005**, *5*, 450–456. [CrossRef]
2. Wang, Y.C.; Zhao, W.X.; Hu, Z.Y.; Zhang, C.; Feng, X.Q.; Zhao, W.; Wang, G.R.; Wang, K.G. Parametric study of the emission spectra and photobleaching time constants of a fluorescent dye in laser induced fluorescence photobleaching anemometer (LIFPA) applications. *Exp. Fluids* **2019**, *60*, 106. [CrossRef]
3. Hoebe, R.A.; van Der Voort, H.T.M.; Stap, J.; van Noorden, C.J.F.; Manders, E.M.M. Quantitative determination of the reduction of phototoxicity and photobleaching by controlled light exposure microscopy. *J. Microsc.* **2008**, *231*, 9–20. [CrossRef]
4. Kuang, C.F.; Yang, F.; Zhao, W.; Wang, G.R. Study of the Rise Time in Electroosmotic Flow within a Microcapillary. *Anal. Chem.* **2009**, *81*, 6590–6595. [CrossRef]
5. Zhao, W.; Yang, F.; Khan, J.; Reifsnider, K.; Wang, G.R. Corrections on LIFPA velocity measurements in microchannel with moderate velocity fluctuations. *Exp. Fluids* **2015**, *56*, 39. [CrossRef]
6. Kuang, C.F.; Wang, G.R. A novel far-field nanoscopic velocimetry for nanofluidics. *Lab Chip* **2010**, *10*, 240–245. [CrossRef]
7. Kuang, C.F.; Qiao, R.; Wang, G.R. Ultrafast measurement of transient electroosmotic flow in microfluidics. Microfluid. *Nanofluidics* **2011**, *11*, 353–358. [CrossRef]

8. Zhao, W.; Liu, X.; Yang, F.; Wang, K.G.; Bai, J.T.; Qiao, R.; Wang, G.R. Study of Oscillating Electroosmotic Flows with High Temporal and Spatial Resolution. *Anal. Chem.* **2018**, *90*, 1652–1659. [CrossRef]
9. Hu, Z.Y.; Zhao, T.Y.; Wang, H.X.; Zhao, W.; Wang, K.G.; Bai, J.T.; Wang, G.R. Asymmetric temporal variation of oscillating AC electroosmosis with a steady pressure-driven flow. *Exp. Fluids* **2020**, *61*, 233. [CrossRef]
10. Hu, Z.Y.; Zhao, T.Y.; Zhao, W.; Yang, F.; Wang, H.X.; Wang, K.G.; Bai, J.T.; Wang, G.R. Transition from periodic to chaotic AC electroosmotic flows near electric double layer. *AIChE J.* **2021**, *67*, 17148. [CrossRef]
11. Wang, G.R.; Yang, F.; Zhao, W. There can be turbulence in microfluidics at low Reynolds number. *Lab Chip* **2014**, *14*, 1452–1458. [CrossRef]
12. Wang, G.R.; Yang, F.; Zhao, W. Microelectrokinetic turbulence in microfluidics at low Reynolds number. *Phys. Rev. E* **2016**, *93*, 013106. [CrossRef] [PubMed]
13. Zhao, W.; Yang, F.; Wang, K.G.; Bai, J.T.; Wang, G.R. Rapid mixing by turbulent-like electrokinetic microflow. *Chem. Eng. Sci.* **2017**, *165*, 113–121. [CrossRef]
14. Zhao, W.; Wang, G.R. Scaling of velocity and scalar structure functions in ac electrokinetic turbulence. *Phys. Rev. E* **2017**, *95*, 023111. [CrossRef] [PubMed]
15. Wang, G.R.; Yang, F.; Zhao, W.; Chen, C.P. On micro-electrokinetic scalar turbulence in microfluidics at a low Reynolds number. *Lab Chip* **2016**, *16*, 1030–1038. [CrossRef]
16. Zhao, W.; Yang, F.; Khan, J.; Reifsnider, K.; Wang, G.R. Measurement of velocity fluctuations in microfluidics with simultaneously ultrahigh spatial and temporal resolution. *Exp. Fluids* **2016**, *57*, 11. [CrossRef]
17. Brizitskii, R.V.; Saritskaya, Z.Y. Boundary value and extremal problems for the nonlinear convection-diffusion-reaction equation. *J. Alloy. Compd.* **2015**, *463*, 559–563.
18. Lazarov, R.D.; Tomov, S.Z. *Adaptive Finite Volume Element Method for Convection-Diffusion-Reaction Problems in 3-D*; Nova Science Publishers, Inc.: Hauppauge, NY, USA, 2000.
19. Phongthanapanich, S.; Dechaumphai, P. A characteristic-based finite volume element method for convection-diffusion-reaction equation. *Trans. Can. Soc. Mech. Eng.* **2008**, *32*, 549–559. [CrossRef]
20. Taneja, S.; Rutenberg, A.D. Photobleaching of randomly rotating fluorescently decorated particles. *J. Chem. Phys.* **2017**, *147*, 104105. [CrossRef]
21. Gavrilyuk, S.; Polyutov, S.; Jha, P.C.; Rinkevicius, Z.; Ågren, H.; Gel'Mukhanov, F. Many-Photon Dynamics of Photobleaching. *J. Phys. Chem. A* **2007**, *111*, 11961–11975. [CrossRef]
22. Hintschich, S.I.; Rothe, C.; King, S.M.; Clark, S.; Monkman, A.P. The Complex Excited-state Behavior of a Polyspirobifluorene Derivative: The Role of Spiroconjugation and Mixed Charge Transfer Character on Excited-state Stabilization and Radiative Lifetime. *J. Phys. Chem. B* **2008**, *112*, 16300–16306. [CrossRef] [PubMed]
23. Ehrlich, K.; Kufcsák, A.; Krstajić, N.; Henderson, R.K.; Thomson, R.R.; Tanner, M.G. Fibre optic time-resolved spectroscopy using CMOS-SPAD arrays. In Proceedings of the Optical Fibers and Sensors for Medical Diagnostics and Treatment Applications XVII, San Francisco, CA, USA, 28–29 January 2017; International Society for Optics and Photonics: Bellingham, WA, USA, 2017.
24. Lee, Y.U.; Li, S.L.; Bopp, S.E.; Zhao, J.X.; Nie, Z.Y.; Posner, C.; Yang, S.; Zhang, X.; Zhang, J.; Liu, Z.W. Unprecedented Fluorophore Photostability Enabled by Low-Loss Organic Hyperbolic Materials. *Adv. Mater.* **2021**, *33*, 2006496. [CrossRef] [PubMed]
25. Wang, G.R.; Fiedler, H.E. On High Spatial Resolution Scalar Measurement With LIF. *Exp. Fluids* **2000**, *29*, 265–274. [CrossRef]
26. Lai, M.J.; Pan, R.H.; Zhao, K. Initial Boundary Value Problem for Two-Dimensional Viscous Boussinesq Equations. *Arch. Ration. Mech. Anal.* **2011**, *199*, 739–760. [CrossRef]
27. Verschaeve, J.C.G. Analysis of the lattice Boltzmann Bhatnagar-Gross-Krook no-slip boundary condition: Ways to improve accuracy and stability. *Phys. Rev. E* **2009**, *80*, 036703. [CrossRef]
28. Smistrup, K.; Bu, M.; Wolff, A.; Bruus, H.; Hansen, M.F. Theoretical analysis of a new, efficient microfluidic magnetic bead separator based on magnetic structures on multiple length scales. *Microfluid. Nanofluidics.* **2008**, *4*, 565–573. [CrossRef]
29. Mortensen, N.A.; Okkels, F.; Bruus, H. Reexamination of Hagen-Poiseuille flow: Shape dependence of the hydraulic resistance in microchannels. *Phys. Rev. E* **2005**, *71*, 057301. [CrossRef]

MDPI
St. Alban-Anlage 66
4052 Basel
Switzerland
Tel. +41 61 683 77 34
Fax +41 61 302 89 18
www.mdpi.com

Micromachines Editorial Office
E-mail: micromachines@mdpi.com
www.mdpi.com/journal/micromachines

www.ingramcontent.com/pod-product-compliance
Lightning Source LLC
Chambersburg PA
CBHW062106220926
51850CB00010B/2912